HOW TO BUILD
A LOW-COST LASER

RONALD M. BENREY

HAYDEN BOOK COMPANY, INC.
Rochelle Park, New Jersey

Library of Congress Cataloging in Publication Data

Benrey, Ronald.
  How to build a low-cost laser.

  1. Gas lasers—Amateurs' manuals. I. Title.
TA1695.B46   621.36'63   73—11927
ISBN 0-8104-5934-5

Printed in the United States of America

1	2	3	4	5	6	7	8	9	PRINTING

74	75	76	77	78	79	80	81	82	YEAR

PREFACE

My fascination with the laser began many years ago. In the summer of 1960, as a fledgling technical reporter, I attended the first press conference held by Dr. T. H. Maiman (then of Hughes Aircraft Corporation) to announce his development of the *ruby laser*. The device was clearly revolutionary, thoroughly intriguing, and surprisingly simple! Soon after, I began working to develop a basic ruby laser that could be built by a home experimenter. Unfortunately, costs of the key components remained outlandishly high until 1964, when plans for my do-it-yourself laser appeared in *Popular Science Monthly*.

About a year after the ruby laser's birth, a research team of physicists at the Bell Telephone Laboratories, led by Dr. A. Javan, announced the development of the *gas laser*—an instrument that produces a continuous beam of laser light, rather than the brief—but intense—burst generated by a ruby laser. That first gas laser was an incredibly finicky device; Bell technicians were concerned that the vibration of subway trains running beneath the Manhattan hotel where the announcement was made would disturb the demonstration. That first laser was in marked contrast with the simple, stable gas laser described in this book.

In the beginning, many people called the laser "a solution waiting for a problem," because the laser had few practical applications during the early 1960's. Those days are gone forever! It's hard to name a modern technology—everything from medicine to weapon design to space exploration—where the laser is not playing an important role. And now, at long last, the gas laser has come of age as a safe, practical, and reasonably priced analytical tool for the creative amateur experimenter.

I have planned this text as an introduction to gas laser technology and two of its most significant applications:

- *holography,* the creation of unique three-dimensional image "records"
- the study of classical optical phenomena (through a series of "fun" experiments designed to show how light "works")

The book is structured to present basic theory, experimental technique, and practical information in a logical sequence. Wherever possible, I have minimized complex theory, and emphasized fun; this is as it should be, since this book is for the hobbyist. And I have designed the optical experiments and holography setups with an eye to keeping costs low. The "optical bench," for example, is made out of standard copper plumbing supplies and common hardware.

The most expensive item is, as one might expect, the gas laser. A factory built unit costs about $100. However, you can build your own laser from readily available components, in kit form, for about $80. *All* of the other components and materials cost about $50—including enough special film to make fifty holograms.

I would like to dedicate this book to my son, Andrew, who is always asking me "How does that work?" I hope this book will answer a few of his questions—and a few of yours!

Ronald M. Benrey

Rochester, Michigan

CONTENTS

SAFETY NOTICE

A laser can be used as a light source to perform exciting experiments and projects. You have probably read or heard about powerful pulsed lasers that can weld or drill metal. The low-power, continuous-wave laser described in this book, however, is incapable of such feats. In fact, this laser could be aimed at clothing or skin without doing any damage. However, just as people are cautioned not to look directly into a bright slide projector, they should also be cautioned not to look directly into the laser while it is operating. Staring directly into the laser for an extended period of time could result in eye damage that is similar to sunburn.

Simple safety considerations are given in Chapter 1. Be sure to read them before you build and use your own laser.

1

HOW LASERS WORK

Laser Principles and Construction

The gas laser is a unique—and highly specialized—source of light. The beam of light it produces has three significant characteristics that make it useful in optical experimentation, *holography* (lensless three-dimensional "photography"), and other scientific applications. Although light possessing the first two properties can be produced by some conventional light sources, only laser light possesses all three:

1. Laser light is *monochromatic.* The light waves in the beam are all of the same frequency (and, thus, of the same wavelength). The beam has a single, pure color.
2. The laser beam is highly *collimated.* This means that the beam remains narrow, and can travel long distances, without spreading apart. Optical scientists say that the beam has narrow divergence.
3. The laser light is *coherent.* This means that all the individual light waves making up the beam are in phase—or in step—with each other. We'll talk more about this key property later.

In addition, the laser beam is very intense; the laser is a powerful light source.

The name "laser" is an acronym for **L**ight **A**mplification by **S**timulated **E**mission and **R**adiation. This mercifully shortened

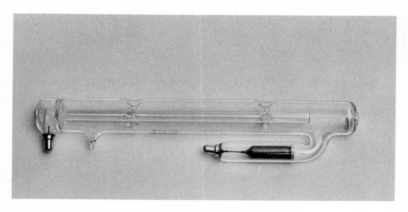

Fig. 1-1.   Gas laser.

mouthful gives a clue about how a laser operates, although the word "amplification" can be confusing. Although the heart of a laser is an amazing "light wave amplifier," the device produces a beam when it begins to *oscillate*. Think of the laser as a special kind of oscillator and you'll understand its operation. And therein lies a fascinating tale.

A gas laser is a surprisingly simple gadget . . . at least its structure is simple. In many ways it is similar to the neon electric signs that light up store windows and marquees. A gas laser (Fig. 1-1) consists of a thin glass tube about 1 foot long, and filled with a low-pressure mixture of helium and neon gases. The gases are blended in a ratio of about one part neon to about six parts helium, and the gas pressure in the tube is about 1/300 sea level atmospheric pressure.

A pair of electrodes—a *cathode* (or negative electrode) and an *anode* (or positive electrode)—are mounted near the ends of the tube (Fig. 1-2). These electrodes are connected to a high-voltage, direct-current (dc) power supply. The gas laser we will assemble and use has a 1,700-volt dc power supply; higher voltages are needed for larger instruments.

The power supply is equipped with a *starting circuit* that momentarily boosts the voltage to about 5,000 volts dc when the laser is turned on. The powerful *electric field* produced between the

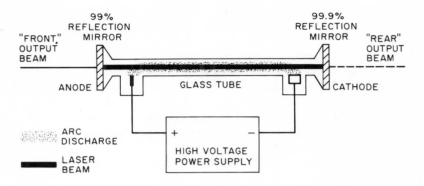

Fig. 1-2. Inside a gas laser.

two electrodes *breaks down* the column of gas, instantly transform-
ing it from a poor conductor of electricity into a relatively good
conductor, thus permitting a continuous electric *glow discharge* to
take place within the glass tube. Simply, this means that a continuous electric current flows between cathode and anode, through the
partially *ionized* column of gas.

When this happens inside a neon sign tube, the tube emits its
characteristic reddish-orange neon glow. This is because the electric
discharge causes countless collisions among the gas atoms that *excites* them to high energy states; as the atoms randomly fall back to
their normal energy states they each emit a bundle of light energy
called a *photon.* Photons can behave both like waves and like
particles, so in the explanation ahead we will sometimes refer to
them as light waves, and sometimes as packets of energy.

You have to be an atomic physicist to really understand this
process, but a useful model—and keep in mind it is a rough model—is
to picture each gas atom as a miniature solar system consisting of a
positively charged "sun" (the *nucleus*) made of *protons* and *neu-
trons,* and a group of negatively charged "planets" (the *electrons*).

When the atom is in its normal energy state, the electrons
whirl close to the nucleus; however, at higher energy states they
move farther away, or spin in more complex patterns around the
nucleus. When the electric discharge excites a particular atom, one or
more of its electrons shifts instantaneously to a higher orbit, and the

atom enters a higher energy state. As the electron(s) snaps back to a normal-energy-state orbit, the sudden shift releases a packet of light energy.

In a conventional neon sign, these up and down energy shifts take place pretty much at random within the glowing gas column. At any given instant some atoms are randomly being excited, while other previously excited atoms are losing energy and emitting light waves. In a gas laser, by contrast, the downshifts from excited state to lower energy level *do not* take place randomly—they are *stimulated.* Here's what happens:

The mixture of gases in the laser tube has been carefully selected so that, when the discharge occurs, a so-called *population inversion* of energy states is created. Simply, this means that more neon atoms are in high energy states than in low energy states.

Note that it is the neon gas that is responsible for "lasing." The helium gas is present to make an efficient population inversion of the neon possible. As the discharge excites the helium atoms, they collide with the neon atoms and transfer energy to the neon atoms, thus raising their energy state.

Because of the population inversion, *stimulated emission* of photons can take place. Suppose that one excited neon atom drops to a lower energy level and emits a light wave. As this light wave speeds past other excited atoms, it stimulates them to emit their photons in a sort of atomic relay race.

This process can be considered a kind of amplification, since one photon entering the column of gas at one end causes other photons to be emitted. Thus more than one photon comes out the other end. The first photon has been amplified by the column of glowing gas.

Now for the magic touch: We add a pair of reflecting mirrors to the ends of the glass tube. These mirrors cause any light waves emitted in the direction of the tube (parallel to the tube) to bounce back and forth along the gas column. This quickly establishes a continuous "lasing" action.

The electric discharge continuously *pumps* the neon gas atoms to higher energy states, and, at the same time, light waves bouncing between the mirrors stimulate the excited atoms into emitting their packets of light energy. As a result, a steady stream of coherent, monochromatic light is generated within the column. Note that the light radiation is coherent because the stimulated atoms emit their light waves in phase—or in step—with the pattern of light waves bouncing back and forth between the mirrors. And the light is monochromatic since each stimulated atom loses the same quantity of energy as it changes state. Thus each emitted photon has the same wavelength, energy, and color.

In effect, the gas laser is a light oscillator that generates a coherent stream of light waves; it can be compared to an electronic oscillator that generates a coherent flow of radio waves. The electronic oscillator has an amplifying vacuum tube or transistor as its heart; the laser uses a column of excited gas atoms as a "light amplifier."

The electronic oscillator uses components that feed-back part of the output signal to the input of the amplifier; this is what sustains oscillation. The laser uses two mirrors that bounce back coherent light waves into the gas column.

Of course laser light bouncing around inside the tube doesn't do us any good—we've got to get some of it outside. To do this, we specify that one of the mirrors be *partially transmitting;* we require that it passes part of the light that strikes it. Specifically, the laser tube's "front" mirror is designed to reflect 99% of the light that hits it, and allow the other 1% to pass through. Thus, a tiny 1% of the laser light generated within the tube passes through the front mirror as a narrow light beam. This is the amazing laser beam!

Properties of Laser Light

We've spoken in general terms about laser operation; now let's get down to specifics. We know that laser light is monochromatic, but what color is it? The helium-neon gas blend we've described above produces a deep-red output beam. Optical scientists characterize the color of a light wave by specifying its wavelength in nanometers (nm). Each nanometer equals one-tenth of a millionth of a centimeter—an exceedingly small distance.

The helium-neon gas laser beam has a wavelength of 633 nm. By contrast, violet light measures about 400 nm (violet is at the other end of the spectrum of visible colors). The human eye can detect a range of light colors ranging from about 400 nm to approximately 700 nm.

As a brief review, Figure 1-3 shows the relationship between *wavelength, frequency,* and path of *propagation* of a light wave. As the wave moves forward through space, its magnetic and electric components alternately reach positive and negative peaks. The variation with time of either the electric or magnetic components can be drawn as a sine-wave plot in which the vertical-axis reading indicates the instantaneous amplitude of the selected component, and the horizontal-axis reading indicates the position in space of the front of the wave.

A complete light wave cycle includes both a positive and a negative portion. Frequency is defined as the number of cycles taking place each second; wavelength is defined as the spacial distance between successive peaks (either positive or negative).

Optical Mirrors

Another point to consider is the design of the mirror. To most of us the term "mirror" conjures up the idea of a silvery metal film deposited on a glass surface. This is the kind of mirror we use every day. Unhappily, a silver-surface mirror is not efficient enough for use

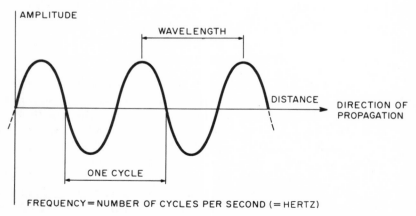

Fig. 1-3.   Frequency and wavelength of a light wave.

in a gas laser. This is because the metal film *absorbs* part of the light that strikes it (often as much as 20%).

Continuous lasing action depends on the continuous rebounding of the coherent chain of light waves off the end mirrors. There are losses inside the glowing column that tend to weaken the coherent beam, so that even after the amplifying effect of the stimulated emission process, a single pass through the column amplifies the light wave only about 2%.

Thus normal mirror losses can't be tolerated. If ordinary silver mirrors were used, the lasing action would never start, since the stimulated emission process can't make up the absorption loss.

As we've said, we plan to take 1% of the laser light energy out of the column as our output laser beam. Obviously this reduces the quantity of light in the column. In fact, it leaves only 1% "extra" light energy in the column, since the output beam and overall losses can not exceed the 2% amplification factor of the column (or else lasing will cease). Thus the mirrors must be highly efficient; they cannot absorb more than a tiny fraction of light striking their surfaces.

The type of mirror used is made by applying many very thin layers (often as many as thirteen) of *dielectric* material to a glass surface. The thickness of each layer in the "sandwich" is carefully controlled to produce a mirror that reflects light of a single color only; the color, of course, is 633 nm deep red. Dielectric materials are substances characterized by low electrical conductivity, although in this application it is their optical properties with which we are concerned. We'll simply say that they are transparent crystalline substances that can be deposited in ultra-thin layers (each a fraction of a thousandth of an inch thick) on a glass surface.

Thin-film mirrors (each dielectric layer is usually called a film) utilize the phenomenon of lightwave inteference, a topic we'll discuss in Chapter 5, to reflect light of specific color. You observe this type of mirror at work when you look at an oil slick on a wet pavement. The slick looks like a grounded rainbow because different areas of the slick reflect different color light. This is because each area of the slick is actually a thin film of specific thickness. Each thickness zone reflects a particular color of light, creating the multicolor display.

Thin-film mirrors are exceedingly efficient: When light of the specific color for which they are designed strikes them, they absorb only a tiny fraction of the light. Depending on the number of layers used, a thin-film mirror will reflect well over 99% of the light striking it. And a thin-film mirror can also be designed to transmit (let pass) a small fraction of the light striking it. Thus, both the front and rear mirrors of a gas laser can be thin-film units.

A Practical Laser

As you might expect, the simple "theoretical" gas laser we've talked about requires a few modifications to make it into a practical device. We will illustrate these changes by discussing the laser we'll be using to perform the practical experiments and as a holographic "light source": the model ML-610 laser manufactured by Metrologic Instruments Inc., 143 Harding Avenue, Bellmawr, N.J. 08030. Similar instruments are also available from Edmund Scientific Co., Barrington, N.J. and Sargent-Welch, Spokie, Illinois. The laser tube and power supply used in this instrument are available separately only from Metalogic for those hobbyists who wish to assemble their own gas laser (Chapter 2).

Consider the mirror geometry first. If perfectly flat mirrors are used at the tube ends, the mirrors must be positioned perfectly parallel for lasing to take place. Obviously, only with parallel mirrors will the beam bounce back and forth *within* the tube. Any misalignment—due to vibration, or flexing of the tube, or manufacturing tolerance—will deflect the beam out through the tube walls, and kill the lasing action.

This type of gas laser can be built, but it is expensive to manufacture and hyper-finicky to set up and to use. A solution is to use a *semi-confocal* mirror arrangement (Fig. 1-4). Here the rear mirror is flat, while the front mirror is slightly curved. The curvature of the front mirror tends to reflect light back toward the rear mirror (and, thus, through the glowing gas column).

Notice that the combination of flat and curved mirror surfaces produces a cone-shaped beam inside the column of gas. Because a parallel output beam is desired, the front surface of the front mirror

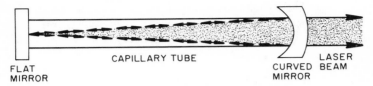

FLAT   CAPILLARY TUBE   CURVED   LASER BEAM
MIRROR   MIRROR

Fig. 1-4. Semi-confocal mirror arrangement.

is ground into the shape of a convex lens that bends the cone-shaped beam into a parallel beam.

The inside surface of the front mirror is coated to make it 99% reflective, while the inner surface of the rear mirror is made 99.9% reflective. Note that there is a cone-shaped rearward output beam that passes through the 99.9% mirror . . . but it is only 10% as intense as the frontward beam.

For structural reasons, a thin glass "capillary" tube by itself is an impractical device; slight flexing will crack it. Thus our practical laser tube has an outer glass tube to support the inner tube and the two end mirrors. This large outer tube rigidly aligns the mirrors and capillary tube, and provides additional strength (to make the tube easy to handle). In addition, the outer tube serves as a reservoir for extra helium-neon gas mixture: During lasing, neon gas is slowly absorbed by the cathode.

The power supply deserves comment too—it is more than just a simple dc source. As we've said, when it is first turned on it produces a high-voltage starting pulse to break down the column of gas, and initiate the glow discharge. As soon as the glow discharge is established, the power-supply voltage drops to a nominal 1,700 volts dc. A chain of resistors in the power supply, wired in series with the anode electrode, limits the current flow through the laser tube from exceeding safe limits.

If you look down at the top of the glowing laser tube (after observing the safety precautions outlined at the end of this chapter) you will see a bright (pinkish-orange) glow in the capillary tube, and somewhat dimmer glowing regions near the anode and cathode. The glow is similar to—but a good deal paler—than the glow from a neon sign.

Note that only the laser output beam is deep red. These other glows are produced by other *transitions* of atomic energy states within the glow discharge.

Coherence

The laser is unique because it is a source of coherent light. Coherence, as we have said, means that the individual light waves making up the laser beam are in phase, or in step, with each other *both in time and in space.*

A conventional light source—an ordinary incandescent light bulb, for example—is a source of non-coherent light waves. As electric current passes through the filament, the filament heats to a high temperature. The resulting atomic collisions among the in-motion atoms excites many of them to higher energy states. As they randomly return to lower energy levels, they emit photons.

The filament glows white because it simultaneously emits light waves of all different colors that blend together to create white light. The light is non-coherent because the up-and-down energy shifts of the various excited atoms are independent of each other.

The laser is a coherent light source because the wave of light sweeping back and forth through the glowing gas column (between the mirrors) stimulates the excited atoms into emitting their waves in step with the wave. It's a kind of "bandwagon" effect.

As a rough analogy to help explain the concept, imagine someone dropping pebbles into a still pond. First assume that he drops the pebbles at random, and they fall at different times into different spots in the pond. Picture the kaleidoscopic pattern of ripples that would result. Each pebble would create its own set of circular ripples that move out from the striking point. The various ripples would interact to create a meaningless jumble; the sets of ripples are non-coherent.

But next assume that the man were careful to drop every pebble in precisely the same spot. And further assume that he timed the arrival of each pebble perfectly, so that it hit the water in step with the existing ripple pattern. Now the ripple pattern will be a neat series of concentric circles. Every ripple will be in phase with the one ahead and the one behind. The ripples are coherent in terms of time measurement.

In a gas laser, the sweeping light wave stimulates countless neon atoms distributed across the column's cross-section. Each of these atoms emits a photon in step with the moving wave, and consequently these emitted photons are in step with other. Thus a coherent beam is produced.

It should be noted that coherent radiation is only possible when all the waves have the same wavelength. Thus a coherent light source is monochromatic.

Because the laser beam is a coherent light source, it can be used to demonstrate optical *inteference* phenomena . . . including holography. In these experiments the laser beam is split into two parts, and these parts are then brought together in such a way that the light waves interfere with each other. We'll discuss how and why in Chapters 5 and 6; for now we'll simply say that *only* two *coherent* light beams will demonstrate inteference effects.

Note, however, that the laser is not a completely perfect coherent light source. There are unavoidable gremlins at work within the gas column that throw off the emitted wave wavelengths of some of the neon atoms. As a result, the laser beam is truly coherent for only a short distance past the front of the tube. This "coherence length" varies from laser to laser. The unit we will work with has a coherence length of a few feet, typically about 3 feet. Thus all of our experimental equipment will be placed within 3 feet of the laser.

We must briefly mention that many other kinds of gas lasers have been designed, using different blends of gas(es). These operate in much the same way as the unit described above, although many use radio-frequency excitation of the gas atoms instead of an electric discharge. Here a small radio transmitter generates radio-frequency signals that are coupled to the gas column by an induction coil. The radio-frequency energy is absorbed by the gas, pushing the atoms to higher energy states.

Safety Considerations

It is a well–publicized fact that some lasers project a beam powerful enough to damage eyesight. The laser described in this book cannot: It produces only about 1/1000 watt of optical power, enough to perform experiments easily, but well below dangerous levels.

However, the beam is still exceptionally bright. And we urge you not to look directly at the beam for the same reasons you don't look directly into a powerful movie projector or an arc light: Essentially, you'll find it uncomfortably bright and you will squint. Similarly, don't look at the reflection of the beam from a shiny metal, or glass, surface.

If you assemble your own laser, or if you look inside a commercially built instrument, keep in mind that when the device is plugged into a wall outlet there are several terminals carrying 117 volts of alternating current (ac), and under certain circumstances *117 volts ac can be lethal.* Be careful not to touch these terminals!

The 1,700-volt power supply can deliver a nasty jolt if you touch its output terminals, or simultaneously touch both laser tube electrodes. Watch what you touch when the laser is plugged in. Better yet, make it a practice *never* to open the laser case while the power cord is connected to an outlet. Finally, *never* leave the laser unattended while it is operating.

2

HOW TO BUILD
YOUR OWN LASER

It takes a physicist to fully explain how a gas laser operates, but you can assemble your own gas laser in a few hours, simply by cobbling together three easy-to-handle "building blocks:"

- The laser tube (Fig. 1-1).
- The power supply module (Fig. 2-1).
- The enclosure (including pilot light and power switch) (Fig. 2-2).

The laser tube is available fully assembled and tested from Metrologic Instruments, Inc., 143 Harding Avenue, Bellmawr, New Jersey 08030. The model number is MT-705; the tube costs $45.00. You must purchase a pre-assembled tube; it is impractical to build your own from scratch.

The power supply module is available fully assembled from Metrologic (model 60-146; $23.50) or in kit form (model 60-143; $18.50). Or you can follow the schematic diagram (Fig. 2-3) and parts list (Table 2-1) and wire up your own power supply module. The component parts are readily available at any broad-line electronics supply house (including most national mail-order electronics supply companies), and the circuit is straightforward. Use a piece of perforated phenolic chassis board to hold the components (except the transformer); use push-in terminals as wiring and soldering points.

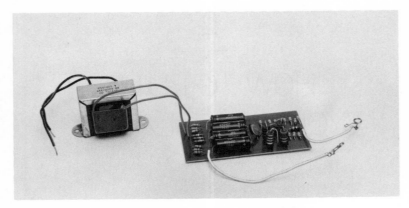

Fig. 2-1.   Laser power supply module.

Fig. 2-2.   Laser enclosure.

Probably, though, most experimenters will find it much simpler—and possibly less expensive—to purchase the ready-to-wire kit,
which includes a printed circuit board to hold and to interconnect
the components. Therefore we will only discuss assembly of the kit
later in this chapter. Experienced electronics buffs will be able to

duplicate the power supply without additional instruction, by working from the schematic diagram.

The laser enclosure is available in kit form from Metrologic, or you can assemble the alternate enclosure described later in this chapter; it is designed to work with the low-cost *optical bench* illustrated in Chapter 4.

Metrologic also offers a complete laser kit (model MK-610) for $83.00 (postpaid) that includes the tube, power supply kit, and enclosure kit. When you have completed assembly, the finished product is equivalent to Metrologic's model ML-610 laser ($87.00— completely assembled).

Also note that any helium-neon gas laser of similiar characteristics can be used to make holograms and to carry out the optical experiments. The ML-310/MK-310 laser produces approximately 0.5 milliwatt of optical power. Its output is randomly polarized, and its output beam diameter (close to the laser) is approximately 1.2 millimeters (mm). As noted earlier, the beam is bright red (633 nm). Several other firms manufacture low-cost gas lasers that will work equally well (although they are not as readily available to the amateur experimenter).

The Power Supply

The heart of the power supply circuit (Fig. 2-3) is a familiar *voltage doubler* configuration built around diode rectifiers D1 through D3 and D4 through D6, and electrolytic capacitors C1 through C4. The transformer, T, converts 120 volts ac line voltage to about 750 volts ac. Rectifiers D1, D2, and D3 charge capacitors C1 and C2 positively with respect to the yellow "common" transformer lead, while D4, D5, and D6 charge C3 and C4 negatively with respect to the common lead. When the laser tube is connected to the supply, the net voltage across the stack of the four capacitors (C1 through C4) is approximately 1,700 volts dc. The chain of four resistors (R6 through R9) acts to limit the current flow through the laser tube to safe limits (approximately 5 milliamperes dc).

The "starting circuit" (the nine rectifiers, D7 through D15, and capacitors C5, C6, and C7) is actually a voltage-multiplier

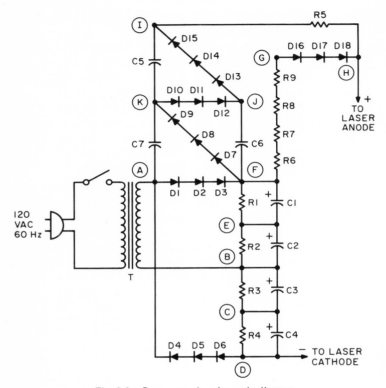

Fig. 2-3. Power supply schematic diagram.

configuration that delivers a series of high-voltage pulses to the laser tube in order to break down the gas column, and start the arc discharge. Diode rectifiers D16, D17, and D18 isolate the pulse circuit from the electrolytic capacitor stack—the capacitors would simply "short circuit" the pulses. Once the laser tube starts to glow, the starting pulses are automatically "swamped out." The relatively low resistance of the arc, in comparison with the high value of R5 (the resistor in series with the pulse circuit) makes the pulse amplitude negligible.

Resistors R1 through R4 serve as "bleeder resistors." They discharge the four electrolytic capacitors when you turn the laser off. Without them the capacitors would hold a considerable charge

Table 2-1. Power Supply Parts List*

Transformer T—this is a special transformer made for Metrologic, and no commercial equivalent exists. A unit that will work (but is large and relatively expensive) is the Stancor PC-8406; this unit is rated as follows:

Secondary windings: #1   650 volts ac at 40 milliamperes
                       #2   6.3 volts ac at 2 amperes
                       #3   5.0 volts ac at 2 amperes

The low-voltage secondary windings are meant to power tube filaments; they are not used in the laser circuit. This transformer is much larger than the Metrologic unit since it has a much greater power-handling capacity.

Resistors R1, R2, R3, R4—1,000,000-ohm, ½-watt, carbon resistors
Resistor R-5—3,900,000-ohm, ½-watt, carbon resistor
Resistors R6, R7, R8, R9—33,000-ohm, 2-watt, carbon resistors
Capacitors C1, C2, C3, C4—5-mfd; 450-volt, electrolytic capacitors
Capacitors C5, C6, C7—0.001-mfd; 2,000-volt, disk capacitors
Diodes D1 through D18—1,000-volt, 1-ampere (or higher), silicon rectifiers (The low-cost Mallory M2.5A is a good choice, but equivalent devices are available from many manufacturers.)

Miniboxes—Bud CU2109A or equivalent (8 X 6 X 3½ in.) and Bud CU2114A or equivalent (12 X 2½ X 2¼ in.)
Miscellaneous—3-wire (grounding-type) line cord set; high-voltage insulated test-prod cable (see text); rubber grommets; 5-terminal, center-terminal ground, terminal strip; SPST toggle switch; "Pop" rivets; hardware; etc.

Note: The components listed are *not* identical to those supplied in the Metrologic 60-143 power supply kit. They are equivalent parts that are available from most broad-line electronics distributors.

---

long after the power was disconnected, introducing a potential shock hazard if you ever opened the enclosure to work on any components.

The power-supply kit takes about 2 hours to assemble. Start by carefully unpacking the plastic bags (Fig. 2-4) that hold the kit components. Segregate the different parts in muffin tins or empty egg cartons, and double-check each item against the parts list. Make sure you know what each component is, and be sure that you know how to identify the polarity marking symbols on the electrolytic capacitors and the diode rectifiers: these components must be installed in the proper "direction" (or orientation) on the board, or they may be damaged (and the power supply will not operate).

Follow the detailed instruction sheet supplied with the kit; it specifies the proper sequence of mounting and soldering the various components. Work carefully, with a hot soldering iron, when you

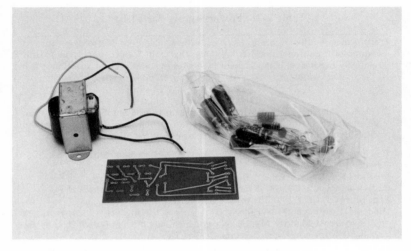

Fig. 2-4.   Power supply kit.

solder component leads to foil areas on the printed circuit board. A
50- or 75-watt "pencil" soldering iron is perfect for the job; do not
use a large iron designed for metal working. The great quantity of
heat produced by this type of unit may cause the bonded copper foil
to peel away from the phenolic backing board.

The Enclosure

If you purchase a Metrologic enclosure kit (or the complete
laser kit) you will receive a professionally crafted enclosure. Follow
the instructions supplied with the kit; assembly should take under 1
hour.

The author's enclosure is built from two readily available
aluminum "miniboxes." (See Power Supply Parts List in Table 2-1
and Fig. 2-5). The laser tube is housed in the long, slender minibox;
the power supply board, transformer, and power switch mount in
the rectangular, squat minibox. The boxes are riveted together so
that the power-supply minibox serves as a support for the laser tube
box.

Note that the pilot lamp (which is included in the commercial
laser) is not installed in our enclosure. This is an intentional omis-

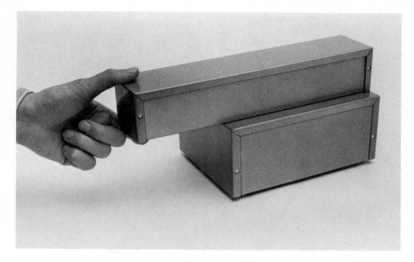

Fig. 2-5.   Minibox enclosures.

Fig. 2-6.   Location of rivets and clearance holes on miniboxes.

Fig. 2-7.   Location of components in minibox.

sion: During holographic experiments, the pilot lamp would be an intolerable source of red light that would quickly fog the exposed film. This point is an important one: If you use the standard Metrologic enclosure for your laser, cover the pilot lamp with black opaque paper when you begin making holograms.

Start to build the enclosure by riveting together (using "Pop" rivets) the top half of the power supply minibox and the bottom half of the laser tube minibox. Use four rivets, as shown in Figure 2-6. Drill two 1/2-in.-diameter holes in the location shown in Figure 2-6 to provide clearance holes for the passage of the leads joining the laser tube and the power supply. Place 3/8-in. inside-diameter rubber grommets in the holes to prevent rough edges from chafing the leads. Drill three 5/32-in. diameter holes around the periphery of the circuit board; these will be used to mount the board in the minibox.

Mount the transformer, the power switch, and the terminal strip in the bottom half of the lower minibox, in the locations shown in the photos. Drill a 1/2-in. hole for the power cord; place a 3/8-in. inside-diameter rubber grommet in the hole (Fig. 2-7). Mount the power supply board with 6-32 screws and nuts; use 3/4-in. long metal spacers (the ones that accomodate 6-32 screws) to support the board away from the aluminum surface.

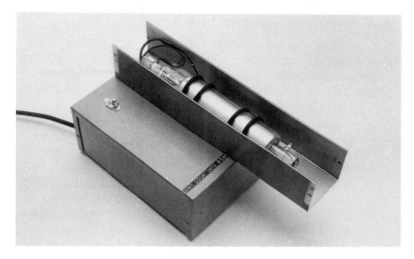

Fig. 2-8.   Laser tube mounting.

The high–voltage lead wire supplied with the kit is not long enough for use in our enclosure. Replace it with two 12-in. lengths of "high–voltage test lead cable" (5,000 volts, or higher, insulation rating). This material is available at any electronics parts house (usually packed in 10-ft rolls).

Mount the laser tube in the bottom half of the upper minibox, as shown in the photograph (Fig. 2-8). Use the rubber "cable clamps" supplied with the tube. Run the high-voltage leads through the grommeted holes, and fit the connector clips over the laser tube terminals. Note that the lead terminating at the negative end of capacitor C4 *must* be connected to the laser tube's *cathode* (the large O-shaped electrode assembly). The positive lead goes to the lone pin-type electrode (the *anode*) housed in the small glass arm attached to the rear of the main body.

After the laser tube is mounted, drill a 1/4-in. beam hole in the "front side" of the top half of the laser tube minibox. Carefully measure the position of the laser tube, so that the hole will line up with the laser tube axis. Complete the enclosure by interconnecting the power cord, power switch, and free transformer leads. Add a set of "press–on" rubber feet ("Scotch Protector Pads"—a standard hardware store item) to the bottom of the power supply minibox.

Care of the Laser Tube

Exercise care when you handle the laser tube. Keep fingers away from the coated end-mirror surfaces: sweat and finger print marks can damage the mirrors. Use a soft "lens brush" (available at photo shops) to brush dirt or dust off the mirrors (Fig. 2-9); do *not* use a cloth or "lens tissue."

Although the tube is well made, it will break if dropped—just like most glass objects. A particularly vulnerable spot is the sealed "nipple" on the end of the tube that was used to fill the tube with gas during manufacture: don't strike or stress this appendage.

Similarly, don't overstress the terminals that connect to the internal electrodes. Be especially careful when you install the connector clips on the terminals. Stress will break the metal-to-glass seals, and will ruin the tube.

Lastly, be very certain that you don't wire the tube "backwards" to the power supply: This will damage the tube severely and may ruin it!

Troubleshooting

There's little to go wrong in the laser tube, or the power supply circuit, so the laser will probably flash on the instant you flip

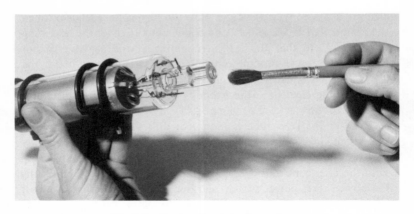

Fig. 2-9. Use lens brush to clean mirrors.

the power switch. Occasionally a new tube (or a tube that has been stored for a long period without use) will take several seconds—or possibly minutes—to come to life. This is caused by gaseous impurities within the tube that are automatically removed, in time, by the "getter" action of the cathode. However, if the laser does not produce a beam after a few minutes, it is likely that the power supply is at fault. To troubleshoot the circuit, you'll need the following equipment:

- A vacuum-tube voltmeter (VTVM) or sensitive volt-ohm-meter equipped with *both* a high-voltage dc measuring range (0–3,000 or 0–5,000 volts dc) and an ohm-meter circuit that covers an overall range of several hundred to several million ohms, in two or three individual, overlapping ranges.
- Five 33,000-ohm, 2-watt, carbon resistors. Wire them in series and keep the interconnecting leads as short as possible.

The power supply schematic diagram (Fig. 2-3) carries a series of circled letters. Each identifies an individual test point that will be used during the troubleshooting procedure.

The first step is to disconnect the line-cord from the ac power socket. Wait 30 seconds for the electrolytic capacitors to discharge, then open the power supply minibox (or the enclosure, if you've assembled the complete kit) and the laser tube minibox. Disconnect the high-voltage leads from the laser tube terminals, and connect them across the five-resistor chain you assembled.

The goal of troubleshooting is to *locate* the faulty or incorrectly wired component(s). We'll assume that you assembled the power supply circuit board carefully, and that you double-checked the polarity of the electrolytic capacitors and diode rectifiers before you mounted them on the circuit board. Further, we'll assume that all solder joints are bright and shiny (dull, granular, joints are "cold soldered," and are poor electrical conductors; reheat any you see, adding a dab of fresh solder).

Troubleshooting is an art, more than a science—an art that takes years of practice to master. However, we can distill the essense of the troubleshooting art into a pair of charts, the Voltage Measurement Chart (Fig. 2-10) and the Resistance Measurement Chart (Fig.

POSITIVE LEAD

	A	B	C	D	E	F	G	H	I	J	K
A	—	NO	NO	NO	NO	NO	NO	NO	NO	NO	NO
B	NO	—	NO	NO	400 / 1,2,3	NO	NO	NO	NO	NO	NO
C	NO	400 / 6,7,8	—	NO	NO	NO	NO	NO	NO	NO	NO
D	NO	NO	400 / 6,9,10	—	NO	NO	NO	900-1,000 v / 12,13	NO	NO	NO
E	NO	NO	NO	NO	—	400 / 1,4,5	NO	NO	NO	NO	NO
F	NO	NO	NO	NO	NO	—	NO	NO	NO	NO	NO
G	NO	NO	NO	NO	NO	500-600 v / 11	—	NO	NO	NO	NO
H	NO	NO	NO	NO	NO	NO	NO	—	NO	NO	NO
I	NO	NO	NO	NO	NO	NO	NO	NO	—	NO	NO
J	NO	NO	NO	NO	NO	NO	NO	NO	NO	—	NO
K	NO	NO	NO	NO	NO	NO	NO	NO	NO	NO	—

NEGATIVE LEAD

NO = No reading made.

Fig. 2-10. Voltage Measurement Chart.

POSITIVE LEAD

NEGATIVE LEAD	A	B	C	D	E	F	G	H	I	J	K
A	—	under 1.5K / 14	NO	NO	NO	over 2 MEG / 1	NO	NO	NO	NO	NO
B	under 1.5K / 14	—	1 MEG / 7,8	NO	1 MEG / 2,3	NO	NO	NO	NO	NO	NO
C	NO	1 MEG / 7,8	—	1 MEG / 6,9,10	NO	NO	NO	NO	NO	NO	NO
D	NO	NO	1 MEG / 6,9,10	—	NO	NO	NO	NO	NO	NO	NO
E	NO	1 MEG / 2,3	NO	NO	—	1 MEG / 1,4,5	NO	NO	NO	NO	NO
F	under 200 / 1	NO	NO	NO	1 MEG / 1,4,5	—	NO	NO	NO	NO	over 2 MEG / 15
G	NO	NO	NO	NO	NO	132 K / 20	—	over 500 K / 18	NO	NO	NO
H	NO	NO	NO	NO	NO	NO	under 200 / 18	—	3.9 MEG / 19	NO	NO
I	NO	NO	NO	NO	NO	NO	NO	NO	—	under 200 / 17	NO
J	NO	NO	NO	NO	NO	NO	NO	NO	over 2 MEG / 17	—	under 200 / 16
K	NO	NO	NO	NO	NO	under 200 / 15	NO	NO	NO	over 2 MEG / 16	—

NO = No reading made. *Note:* Allow several seconds for ohmmeter reading to stabilize when making a reading.

Fig. 2-11. Resistance Measurement Chart.

1. Faulty diode—D1,D2,D3
2. Faulty capacitor C2
3. Faulty resistor R2
4. Faulty capacitor C1
5. Faulty resistor R1
6. Faulty diode—D4,D5,D6
7. Faulty capacitor C3
8. Faulty resistor R3
9. Faculty capacitor C4
10. Faulty resistor R4
11. Check other voltages; then check resistors R6,R7,R8, and R9
12. Check other voltages; then check diode rectifiers D16, D17, and D18
13. Faulty laser tube
14. Faulty transformer T
15. Faulty diode—D7,D8,D9
16. Faulty diode—D10,D11,D12
17. Faulty diode—D13,D14,D15
18. Faulty diode—D16,D17,D18
19. Faulty resistor R5
20. Faulty resistor—R6,R7,R8,R9

Fig. 2-12.   Master Troubleshooting Chart.

2-11). The charts are self-explanatory: The "normal reading" in each box corresponds to the value you should read on the voltmeter or ohmmeter when the positive (+) and negative (−) leads of the instrument are contacting the indicated test points. Each pair of test points intersect on the chart to give the "normal" reading as "seen" between those circuit points. No readings are made in the boxes marked "NO."

Make *voltage checks* while the line cord is *connected* to the ac power line; be careful *not* to touch any circuit board terminals with your fingers. Remember that the voltages present can deliver painful—and possibly dangerous—shocks.

However, make *resistance checks* only when the line cord is *disconnected* from the power line. A "live" circuit board will damage your ohmmeter.

Note that each "normal reading" box has a small number in its lower right-hand corner (a few boxes have two or more numbers).

These numbers correspond to different line numbers on the Master Troubleshooting Chart (Fig. 2-12). If you come across an incorrect voltage or resistance reading during your tests, consult the given numbered line in Figure 2-12. The *hint* you will find listed *suggests* which component *may* be at fault.

You can test individual diode rectifiers with your ohmmeter (set it to a moderate resistance range). Essentially, the meter will read a low resistance value (a few hundred ohms) when the positive (+) ohmmeter lead (normally the red lead) is contacting the rectifier's *anode* and the negative (−) lead is contacting the *cathode.* Reversing the leads will cause the ohmmeter to display a high-resistance reading (millions of ohms).

A high-resistance reading with the leads in either orientation means that the rectifier is "open" (disconnected internally). A low-resistance reading in both cases means that the diode is "shorted."

Keep one point in mind as you make your readings: Component tolerances, power line voltage differences, and voltmeter/ohmmeter errors may change the "normal readings" on your power supply as much as 15 or 20% from the values given in the charts. Ignore minor voltage or resistance differences.

3

ALL ABOUT OPTICAL COMPONENTS

When you perform the optic and holographic experiments described later in this book you will work with a variety of optical devices and components. A few—such as *lenses* and *mirrors*—are probably familiar to you; others, such as *diffraction gratings* and *beam splitters,* may seem new and unusual. In this chapter we'll briefly discuss each of these components, zeroing in on what they are, what they do, how they work, and how to handle them.

Lenses

A lens is a piece of glass (or other transparent material; plastic lenses are often used in inexpensive cameras and binoculars) that has been shaped in such a way so as to cause a light beam to *converge* or *diverge* as it passes through the lens. When a *parallel* beam of light (all the light waves in the beam are parallel) strikes a *converging lens,* the beam is reformed—the light waves come together some distance away from the lens at a point called the *focal point.* The distance between the focal point and the center of the lens (assuming the lens is relatively thin) is called the *focal length.* The most common type of converging lens is the so-called *double-convex* or *biconvex* lens illustrated in Figure 3-1. Both of its two surfaces are convex ("bending outward") spherical surfaces; the focal point, focal length, and converging properties are indicated in the diagram. Figure 3-2 shows two other oft-used converging lens shapes: the *plano-convex* (one

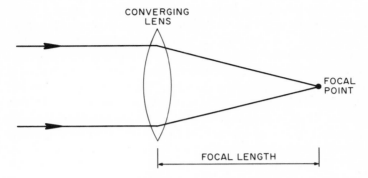

Fig. 3-1. Double convex (biconvex) lens.

PLANO-CONVEX   CONCAVO-CONVEX

Fig. 3-2. Converging lens.

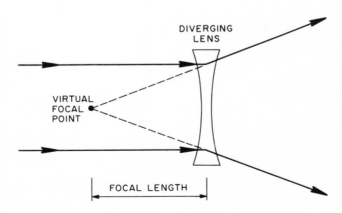

Fig. 3-3. Lens focal point.

flat surface, one convex surface); and the *concavo-convex* (a sharp-ly-curved convex surface on one side, and a gentle concave surface on the other). The latter lens is often called a *positive meniscus lens.*

When a parallel light beam strikes a *diverging* lens the light rays spread apart, and form a cone-shaped beam. Note that the "apex" of the cone is an imaginary point on the "front side" of the lens (see Fig. 3-3). This is the lens's focal point; its focal length is the distance between the point and the center of the lens. Because the focal point is imaginary, physicists indicate the focal length of a diverging lens as a negative number; they simply place a minus sign (-) in front of the focal length specification. Thus in the following chapters we will use a - 8.5-mm diverging lens. The lens illustrated in Figure 3-3 is a *double-concave* lens. Two other diverging lenses are shown in Figure 3-4; the *plano-concave;* and the *convexo-concave* (or *negative menis-cus*) lens.

In later chapters we will use lenses both to converge and diverge light beams. Handle them gently, and keep finger prints off the curved surfaces.

Mirrors

The strict definition of a mirror is "a surface that reflects most of the light falling on it." Actually the common mirrors with which we all are familiar have *two* reflective surfaces: the front glass

PLANO-CONCAVE   CONVEXO-CONCAVE

Fig. 3-4. Diverging lenses.

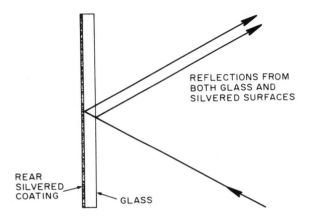

Fig. 3-5.   Reflections from a common mirror.

surface reflects some of the light falling on it, and the rear silvered surface reflects some more. Use a pocket mirror (or any other common mirror) to reflect, a laser beam onto a white screen, and you will actually see *two* spots. One represents the beam's reflection from the front glass surface, the other, the beam's reflection off the silvered surface (Fig. 3-5).

For this reason ordinary mirrors can not be used for holography and optical experimentation. Instead, we use *front-surface* (or "first-surface") mirrors. Here the silver coating has been applied to the front glass surface; the glass merely serves as a "holder" for the shiny metal surface (a perfectly flat holder, we must add). Since there is only one reflective surface, a laser beam is not distorted when it is reflected by a front-surface mirror.

Because they are designed for optical equipment use, and are made with greater precision, front-surface mirrors are considerably more expensive than their familiar cousins, and they are delicate. The exposed silvered surface is easily scratched. Use a soft lens brush—not "lens cleaning tissue"—to whisk dust and dirt off the surface. Also handle these mirrors by the edges only; fingerprints will damage the surface.

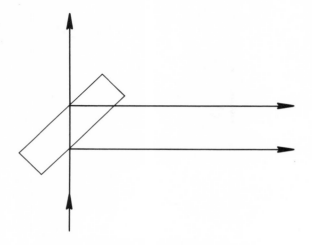

Fig. 3-6. Beam splitter.

Beam Splitter

A beam splitter is nothing more than a small piece of plate glass whose edges are almost perfectly parallel. When the splitter is placed at an angle of 45 degrees to a laser beam (Fig. 3-6), part of the beam reflects sideways off the front surface; another part reflects sideways off the rear surface; and a third part passes through the splitter.

We will use the splitter to create two identical beams when we make holograms: one sideways beam; and one straight-through beam (the second sideways beam will be blocked by a "mask" (see Chapter 8).

Handle the beam splitter with the same care as you would a lens; keep its surfaces clean and free of fingerprints.

"Knife Edge"

We will use the edge of a single-edged razor blade as a "knife edge" to display *straight-edge diffraction phenomena* and as the moving element in the *Foucault Test*. (See Chapter 5 for both applications.)

Polarizing Filters

Polarizing filters are squares of material similar to the material used in sunglasses and photographic polarizing devices. Essentially a polarizing filter consists of two layers of thin protective plastic that sandwich the actual polarizing material. This is a thin sheet of plastic impregnated with countless microscopic crystals of a *doubly refracting* optical material. The crystals are positioned so that their optical axes are all parallel. The mechanism of double-refraction is too involved to be described here; let it suffice it to say that *randomly polarized* light striking the filter is transformed into *plane polarized* light as it passes through. The terms "polarization," "random polarization," and "plane polarization" are all explained in Chapter 5.

The polarizing material you will buy (see Table 4-1 in Chapter 4) is supplied in unmounted sheets. You will find it more convenient to work with if you mount small rectangles of the material in standard 35-mm photographic slide mounts. These mounts are available at any photo supply shop, from a variety of manufacturers. Purchase the kind that are sealed together by running a hot iron around the folded edges.

Single and Double Slits

These are home-made components that will be used to demonstrate two classic optical diffraction phenomena. Make them as follows:

Take two ordinary microscope slides (see Table 4-1) and run each back and forth over a *candle* flame (Fig. 3-7) until you form a uniform coating of thin soot particles on one side of each glass slide. The coatings should be uniformly dark grey—almost black—when you look at a lamp through each slide.

To make the single-slit component, use a double-edge razor blade to scribe a *thin* line across the soot layer on one slide (Fig. 3-8).

To make the double-slit component, press two double-edge razor blades together, and sweep them together across the other slide. You will create two thin, side-by-side lines in the soot coating (Fig. 3-9).

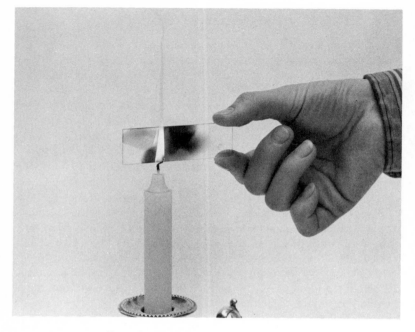

Fig. 3-7. Coating the glass slide with soot.

Obviously the soot coating is very fragile. However, if you damage either of the slit components, you can "remanufacture" it in about 60 seconds!

Diffraction Grating

A diffraction grating is actually an array of side-by-side "slits" that is used to demonstrate diffraction phenomena. The grating you buy (see Table 4-1) will, at first glance, look like a small piece of transparent plastic mounted in a 35-mm slide frame. In reality, you are looking through a "picket fence"-like array of grooves formed in the plastic surface—there are about 13,400 grooves per inch of surface. The grooves are not transparent, but the plastic between the grooves is, creating the *multiple slit* pattern.

Fig. 3-8. Making single slit with razor blade.

Fig. 3-9. Making double slit with two razor blades.

Miscellaneous Components

The *display tank* is simply a small rectangular tropical fish tank—the standard 2- or 2½-gallon size is perfect. The *viewing screen* should be a small sheet of smooth white cardboard.

4

BUILDING AN OPTICAL BENCH

The *optical bench* is the basic tool of experimental optics. It provides a means for supporting the various optical components used in the experiments—and for holography—and makes it possible to shift component position easily and quickly. Professional scientists use heavy—and expensive—benches made of steel or aluminum to guarantee a rock-steady mount for all components; these benches work much like a "monorail" train system. The bench itself is built out of rail–like sections; each component is mounted on a *carrier* assembly that slides along the rail, and can be fixed in any position by tightening a set–screw. Figure 4-1 is a photograph of a professional–style optical bench manufactured by Metrologic Instruments, Inc., Bellmawr, N.J. 08030, that holds a complete holography setup. The firm also sells a variety of optical bench configurations suitable for other optical experiments.

Happily, you can build your own optical bench for a fraction of the cost of a commercially made system. We've designed the bench shown in Figure 4-2 so that it can be assembled out of inexpensive hardware store and stationery shop items. You can duplicate the set-up in the picture for under ten dollars (not counting the cost of the optical components shown).

The bench is built completely out of ½-in. copper plumbing tubing and hardware. Specifically, the *rails* are lengths of ½-in. inner-diameter *stiff* copper tubing; the "joints" are standard *T-couplings* (for ½-in. pipe); and the "feet" are standard ½-in. *elbows*.

Fig. 4-1. Professional optical bench.

All of these items are available at most hardware and all plumbing supply shops. Figure 4-3 gives the dimensions of the eight lengths of copper pipe used to build the bench; you will need an inexpensive *pipe cutter* to cut the various segments from the single, long piece of copper tubing you will buy.

All the "joints" must be soldered, in order to lock the bench into a single rigid unit. Begin by assembling the various pieces of

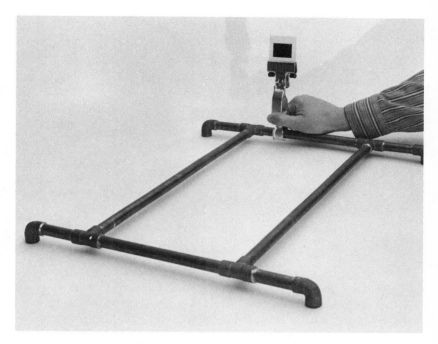

Fig. 4-2. Homemade optical bench.

tubing, elbows, and couplings on a heat-proof surface. The tubing
will hold relatively snugly within the couplings and elbows, so that
the bench will stand together on its own.

You will need a standard propane torch as a heat source, and a
coil of ordinary *rosin core* solder (the kind designed for electronic
wiring use). In turn, heat each "joint" with the torch flame for
several seconds, then touch the tip of the solder to the crack
between the tubing and the elbow or the coupling. If the "joint" is
hot enough, solder will melt, then momentarily form a little ball
atop the crack, and finally flow into the "joint." Remove the flame
as soon as the solder flows, and allow the joint to cool. *Note:* Do not
heat the solder with the flame—heat the joint! The hot joint must be
allowed to melt the solder. Only then will a secure joint be created.
Note also that only a small amount of solder need flow into the joint
to provide a strong bond between tubing and coupling or elbow.

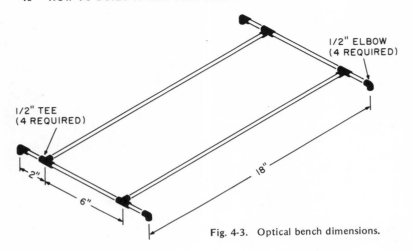

Fig. 4-3. Optical bench dimensions.

The optical component "carriers" are also unusual—they are built atop conventional "50-ampere electrical clamps" (a stock hardware or electrical supply store item). The jaw assembly of this type of clamp fits snugly around ½-in. copper tubing, to hold the clamp (and the optical component mounted on it) upright—yet, a gentle squeeze on the clamp opens the jaws slightly, and permits the carrier to be moved along the bench rails (Fig. 4-4).

In order to perform all ten experiments outlined in Chapter 5, and to make holograms (Chapter 8), you will need all of the optical components and mechanical "assemblies" detailed in the Optical Components Parts List in Table 4-1. This listing itemizes the components, their carriers (when necessary), and general-purpose carriers. The *Part Number* column gives either the Metrologic Instruments (see above) or Edmund Scientific Co. (Barrington, N.J. 08007) parts number for each specific optical component. In the following paragraphs we will describe each of the carriers and "assemblies" that you must build:

Lens Holders

Each lens should be mounted in a simple lens carrier (Fig. 4-5) made of a piece of thin pressed hardboard ("Masonite" or equivalent). You can buy this material at any lumberyard; if you wish, heavy cardboard can be used instead. Cut an off-center location hole that is slightly smaller than the lens diameter, and cement the hole in place with a thin "ring" of epoxy cement. The off-center lens location permits vertical and horizontal adjustment of lens position when the lens holder is mounted in the *carrier/clip assembly* described below.

Carrier/Clip Assemblies

These are the basic carrier devices for the optical bench; they provide a means of mounting the *polarizing screens,* the *diffraction grating,* the *slit components,* the *knife edges,* and the *viewing screen.* Build six of these devices (Fig. 4-6)—three "raised" units, and three

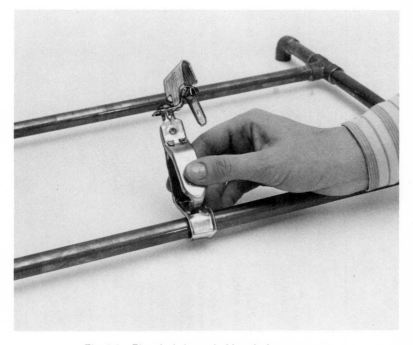

Fig. 4-4. Electrical clamps hold optical components.

Table 4-1. Optical Components Parts List

Component	Number Required	Part Number of Optical Portion	Description of Other Elements and Comments
Carrier/clamp or carrier/clip assembly	6	—	Follow directions in text; use only top-quality electrical clamps as described. The large paper clamps should be heavy-duty units.
-8.5-mm focal length, double-concave lens	1	Metrologic 60-742 Edmund 94,441	Mount in hardboard mount, as described in text. Metrologic unit is premounted.
36-mm focal length, double-convex lens (14.5-mm diameter)	1	Metrologic 60-744 Edmund 94,236	Mount in hardboard mount, as described in text. Metrologic unit is premounted.
152-mm focal length, plano-convex lens (31-mm diameter)	1	Edmund 94,102	Mount in hardboard mount, as described in text.
Diverging lens assembly	2	Metrologic 60-801	For use with beam-splitter assembly described below.
Beam splitter assembly	1	Edmund 41,264 beam splitter	Assemble as described in text; other parts include two diverging lens assemblies (above); tripod head assembly; electrical clamp; angle brackets; miscellaneous hardware; epoxy cement.

Item	Qty	Part / Source	Description
Mirror holder assembly	2	Edmund 40,040 front surface mirror (51 × 76 mm)	Assemble as described in text; other parts include tripod head assembly; 3-in. square piece of hardboard; electrical clamp; miscellaneous hardware; epoxy cement.
Film holder assembly	1	—	Assemble as described in text; parts include tripod head assembly; 3-in. square of hardboard; electrical clamp; miscellaneous hardware; epoxy cement; home-made "film edge retainer."
Subject stand	1	—	Assemble as described in text; make from Bud CU2103A aluminum Minibox (4 × 2 × 2¼ in.); 4 × 2 in. piece of hardboard.
Display tank	1	—	Small rectangular fish tank
Polarizing filter	2	Metrologic 60-722 Edmund 60,637 (Cut sheet of film to size.)	Mount in 35-mm slide mounts as described in text.
Diffraction grating	1	Edmund 40,272	
Sample hologram	?	Metrologic 60-624 Edmund 41,090	*Note:* You can use this sample hologram, if you wish, for experimental viewing before you make your own hologram.

Miscellaneous components: Single-edge razor blades; microscope slides (Edmund P-40,001); accurate ruler; white cardboard; holography film (see Chapter 7); photographic developing chemicals (see Chapter 7); tape.

Fig. 4-5. Lens holders.

"non-raised" units. Basically, each consists of a large office *clamp-type* paper clip mounted atop a 50-ampere electrical clamp. The "raised" units have a spacer rod mounted between the clip and the clamp to raise the clip to a greater height; the "non-raised" units have the paper clip mounted directly to the clamp.

Depending upon the tools and facilities at your disposal, you can bolt, cement (with epoxy), braze, etc., the units together—any firm mounting technique is acceptable. Make sure, though, that the paper clip is positioned so that any flat card-like object held in the slip is oriented perpendicular to the bench rail and to the table top when the electrical clamp is upright.

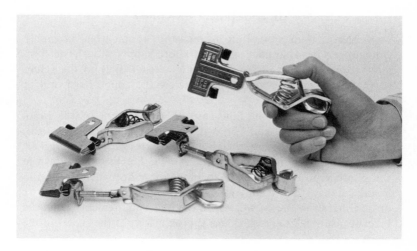

Fig. 4-6. Carrier assemblies.

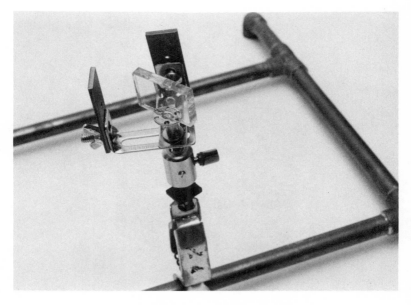

Fig. 4-7. Beam splitter assembly with tripod head.

Beam-Splitter Assembly

This is built around an inexpensive "tripod head" assembly (tilt-and-pan head) available at most photo supply shops. The complete gizmo is shown in Fig. 4-7. Solder a ¼-20 bolt to the clamp (it must be perfectly vertical) and screw the tripod head unit in place.

Cement a ¼-20 machine nut to the bottom of the beam splitter (use epoxy), and allow it to harden in place over night. Mount the two diverging lens assemblies on long-leg angle brackets with 6-32 machine bolts and thumb screws. The brackets shown in the photograph were designed to mount a radio under a car dashboard; however, any readily available brackets (from your junk box or local hardware store) will work; the only requirement is that the bracket position its diverging lens approximately at the center of the beam splitter.

Fig. 4-8. Machine nut cemented to pressed board.

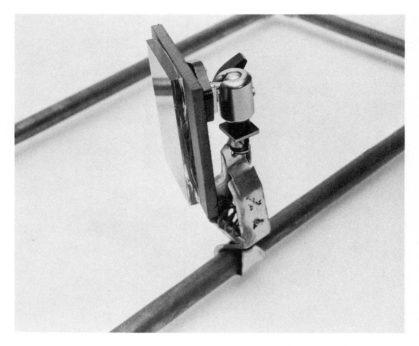

Fig. 4-9. Mirror fastened to stud of tripod head.

Finally, assemble the device by slipping the brackets over the
¼-20 stud projecting from the top of the tripod head, and screwing
the beam splitter in place. Before you tighten the adjusting knob on
the tripod head, be sure that the angle brackets are at right angles to
each other, and that the beam splitter is positioned at a 45-degree
angle to the brackets. The thumb-screws make it easy to adjust the
height and left-right positioning of each diverging lens; the tilt-head
enables you to adjust the orientation of the entire assembly.

Mirror Holders

It is necessary to be able to adjust mirror orientation in all
three dimensions. This capability is achieved, once again, through
the use of adjustable tripod heads. As above, solder a ¼-20 machine

Fig. 4-10.   Film holder made from Scotch electric cable clamps.

bolt to the top of the electrical clamp; mount the tripod-head assembly atop the bolt. The mirror itself is cemented to a small square of pressed board. In turn, cement (with epoxy) a ¼-20 machine nut to the back of the square (Fig. 4-8); the mirror mount fastens directly to the tripod-head stud (Fig. 4-9).

Build two mirror holder assemblies: you will use one to study the basic law of reflection; you will use both to make holograms.

Film Holder

This gadget will hold the film when you make a hologram. It is a carbon-copy of the mirror holder described above, except that the mirror is replaced by a "film edge retainer." You can make the retainer out of thin plastic or metal stock; or, you can (as shown in

Fig. 4-11. Film holder fastened to tripod head.

Figs. 4-10 and 4-11) cement several "Scotch Electric Cable Clamps" (a hardware store item) in place on the face of the hardboard square to make a simple retainer. Use a "sacrificial" piece of film (the same one you will need in Chapter 7) as a template for sizing the retainer; it should hold the film firmly, without buckling it.

Subject Stand and Screen Holder

These components are shown in Figure 4-12. The subject stand (that will hold the subject during holography) is made by cementing a piece of hardboard (2 × 4 in.) atop a 4-in.-tall aluminum minibox (any other 4-in.-high box will work as well). Add four press-on

Fig. 4-12. Subject stand and screen holder.

rubber feet to the bottom of the box. Spray the top of the hardboard rectangle with flat black paint.

The screen holder is simply a office-style paper clamp with its
two "levers" bent backwards to form "feet." This provides you with
an alternate means of supporting a viewing screen: the holder sits on
top of the subject stand, which can be placed anywhere on your
work surface. (For most optical experiments you will mount the
viewing screen atop one of the carrier clamps; however, the interferometer experiment requires a "sideways" viewing screen.)

Important: Several experiments you will perform—particularly
the interferometer and holography experiments—*demand* that all
optical components be held securely in place. Thus, make sure that
the electrical clamps you purchase are capable of gripping *tightly* to

½-in. copper tubing. Be especially wary of low-cost imported clamps; their jaw alignment may be poor, and, as a result, they may "rock" or slip when used to support heavy assemblies such as the beam splitter.

5

LASER EXPERIMENTS

Modern optics is an extremely mathematical science. It must be in order to deal with the complex nature of light, and with the behavior of light as it travels through assorted materials and devices. Thus, this chapter stresses the *observable* characteristics of several optical phenomena, rather than their scientific justification. Happily this approach does nothing to diminish the sheer fascination of the ten "experiments" you will perform as you work your way through the following pages.

Each experiment is described as simply as possible; the description is based on a simple sketch of the optical layout (a sketch is far more understandable than a photograph). Virtually all of the experiments should be done in a darkened room, since the effects (such as diffraction patterns) may be dim. You may find a magnifying glass handy for examining some of the patterns.

If possible, select a rock-steady work surface to hold the optical bench—this is a primary requirement when you set up the *interferometer* (Experiment #10). If you live in a multi-story house, try to work in the basement or on the first floor (this minimizes inevitable house vibration). And set your equipment up on the most rigid—and heaviest—table you can find; a large dining room table is often a good choice.

To minimize unusually severe vibration (such as you might experience in an old house, or a home near a subway system or industrial park) build a "floating table:" Find a large, *heavy* "slab" (a piece of thick plywood, half of an old door, a glass table top, will all do) about the size of a dining room table top. Bolt or cement the optical bench to the center of the slab. "Float" the slab atop a large table with several pieces of foam rubber or polyurethane foam.

One final suggestion: Be sure to keep all of your optical components spotlessly clean; keep finger tips off lenses, polarizing material, diffraction gratings, and glass surfaces.

Experiment #1: Examining the Laser Beam

Two of the laser beam's significant characteristics are readily apparent to the eye: Its pencil-lead-like thinness, and its high intensity.

Aim the laser at a distant wall (30 or more feet away, if possible). Note that the beam projects a relatively small, intensely bright spot on the wall (probably smaller than a dime). The small spread of the laser beam—scientists call it *low beam divergence*—makes possible many laser applications, such as laser-guided weapons, and long-distance laser communication. Professional-quality lasers, costing *hundreds* of times more than your unit, have even less beam divergence.

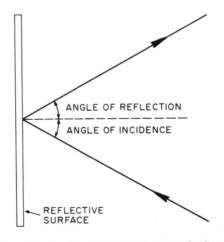

ANGLE OF REFLECTION

ANGLE OF INCIDENCE

REFLECTIVE SURFACE

Fig. 5-1. Angle of incidence equals angle of reflection.

The laser beam (like any other light beam) is invisible as it travels through the air; you see "it" only when it strikes an object, such as a wall, or is "scattered" off tiny particles in the air (thus you will see the beam coursing through dust-laden air). You can make the beam visible by blowing smoke in its path; an even simpler method is to spray almost any aerosol product across the beam. TV turner spray cleaner (available at any electronics supply shop) is ideal, because it won't create a hard-to-clean mess on your table (and on surrounding optics). (*Note:* Be sure to provide adequate ventilation.) For long-duration beam observation, use a water-filled *display tank* (small fish tank). Add a few drops of milk to the water, to provide the light-scattering medium.

If you have a photographic light meter, place it in the beam's path, so that the beam strikes the photosensitive element. You will note a subtantial meter reading, even though only a small part of the total light-sensitive area is being struck (this is true even if your meter has a cadmium sulphide or cadmium selenide photoconductive cell).

Observe the pure color of the beam: You would need a spectrometer to prove the point, but the beam clearly looks like a single-wavelength phenomenon. Many observers consider the color "different" or "unique"—actually, it is merely *unusual:* We rarely see pure, undiluted, color.

Experiment #2: The Law of Reflection

Reflection is probably the most commonly observed optical phenomenon. Each day we see light reflected from countless surfaces, including glass, metal objects, liquid films, and, of course, mirrors.

The basic law of reflection (Fig. 5-1) is familiar to every schoolboy:

angle of incidence = angle of reflection

The narrow laser beam is an ideal "tool" for demonstrating the validity of the law. Mount one mirror holder on the optical bench, and position the laser so that it projects a horizontal beam parallel to one of the long bench rails (in this, and other, experiments, you will have to raise the laser's height by propping it on one or two books,

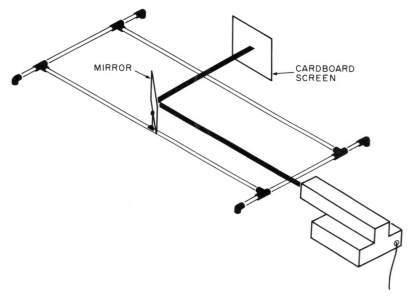

MIRROR

CARDBOARD
SCREEN

Fig. 5-2. Setup for Experiment 2.

or atop a small stack of wood scraps). The laser beam should strike
the mirror roughly at its center.

Set up a large piece of white cardboard in the location shown
in Figure 5-2. As you rotate the mirror to change the angle of
incidence, note how the reflected beam moves along the cardboard;
the angle of reflection always equals the angle of incidence.

Usually when we think of reflections, we almost automatically
think of silvered glass or metal mirrors. However, reflection will
occur when light waves strike the boundry between any dissimilar
optical materials (the dissimilarity being different *indices of refrac-*
tion, a term you will find defined in any elementary optics text).
Thus a piece of clear glass or plastic will also reflect the laser beam
(since its index of refraction is different from the index of refraction
for air). Note though that only part of the beam is reflected; the
other part passes through the "mirror" (Fig. 5-3). The *beam*
splitter you will use later works much the same way.

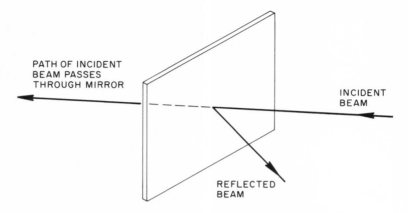

PATH OF INCIDENT
BEAM PASSES
THROUGH MIRROR

INCIDENT
BEAM

REFLECTED
BEAM

Fig. 5-3. Part of beam passes through mirror while other part
is reflected by mirror.

Experiment #3: Basic Lens Optics

The laser beam is an excellent light source for investigating classic *geometrical optics* because it is both bright, and highly *collimated* (it is a parallel beam).

Diverging Lens

Mount your -8.5-mm focal length, double-concave lens on one rail of the optical bench, close to the end nearest the laser. Position the viewing screen about 1 in. (25 mm) away from the lens, as shown in Figure 5-4. Note how the beam is expanded into a larger spot by the lens; slide the viewing screen further away from the lens and the spot will grow larger. If you measure the size of the spot on the screen, with the screen at several different distances from the lens, you will verify that the lens has transformed the parallel laser beam into a cone-shaped beam.

Optical Collimator

A rather simple optical system — one capable of producing a wide parallel beam of light—can be made out of two lenses as shown

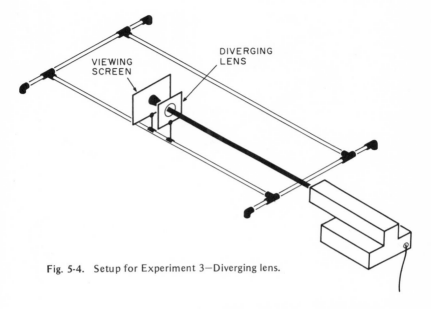

Fig. 5-4. Setup for Experiment 3—Diverging lens.

in Figure 5-5. The negative lens (−8.5 mm) is positioned as above; the second lens (152-mm focal length, plano-convex lens) is placed approximately 3 in. (75 mm) away from the negative lens. Adjust the spacing to produce a parallel—or collimated—beam; you have achieved collimation when the spot on the viewing screen remains constant size as the screen is moved away from the lens system.

Converging Lens

It is possible to use the parallel beam (above) to study the characteristics of your other converging lens (36-mm focal length, double-convex). Mount this lens any convenient distance from the collimator (since the beam is parallel, the distance doesn't matter). Then place the viewing screen approximately 1½ in. away from the converging lens (Fig. 5-6). You should see a tiny spot on the screen, verifying that the lens has converged the parallel beam. Move the screen closer to and farther away from the lens; the spot is at its smallest at the lens's focal point; it grows in size on either side.

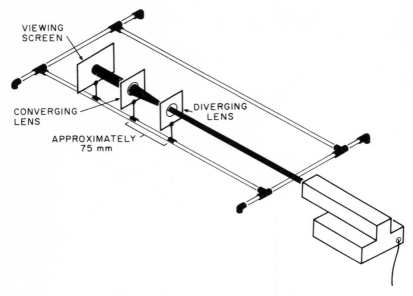

Fig. 5-5. Setup for Experiment 3—Optical collimator.

Experiment #4: Foucault Knife Edge Test

This is a classic test that is frequently used to measure the quality of fine lenses, and it can clear up a typical misconception about the operation of the diaphragm inside a camera lens body.

The experimental setup is the same you used to demonstrate converging lens behavior, above (you will be evaluating the performance of the 36-mm double—convex lens). Begin by sliding the viewing screen to the exact focal point of the lens (point of minimum spot size). Next remove the cardboard viewing screen from the carrier clamp *without disturbing the position of the clamp.* Mount the screen further away from the lens, with another clamp assembly, so that a dime-size spot is projected on the screen.

Now carefully "cut through" the focal point with a single-edge razor blade; use the empty carrier clamp as a guide (Fig. 5-7). If the converging lens is perfect, the visible spot will darken uniformly as the edge passes through the beam. Note that the entire spot is visible

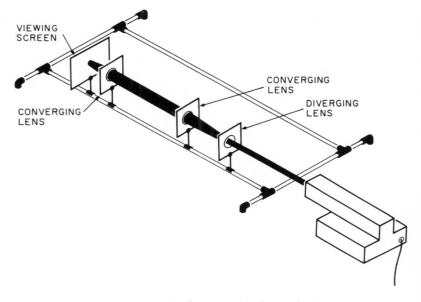

Fig. 5-6.   Setup for Experiment 3—Converging lens.

as it darkens; the blade doesn't chop part of the circle away. Probably, though, you will notice usual light and dark regions in the darkening spot; these indicate assorted imperfections in the lens.

Now for the misconception: The adjustable diaphragm inside a camera lens controls the quantity of light passing through the lens. Most photo buffs understand this, but many find it hard to understand why the diaphram doesn't chop off the outside edges of the image as it closes. The reason is that every part of the lens projects light waves to every part of the image. Thus cutting the flow of light from some parts of the lens (as the razor blade did, or the diaphram does) doesn't cast a "shadow."

Experiment #5: Polarization of Light

Light is one kind of *electromagnetic radiation.* In simple terms, this means that every light wave has both an electric component, and a magnetic component (Fig. 5-8A). These components are

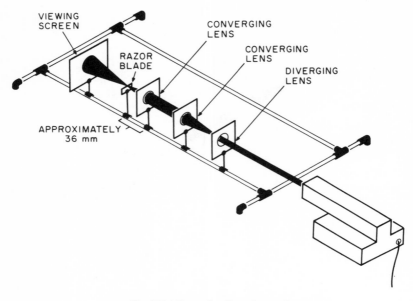

Fig. 5-7. Setup for Experiment 4.

always at right angles to each other, but as the light wave moves through space the components can point in all directions perpendicular to the path of motion (Fig. 5-8B).

Polarization is a term used to describe the orientation of the components (specifically the orientation of the electric component, since it is responsible for the optical effects we are studying). A light source is said to produce *randomly polarized light* when its beam contains waves that have components pointing in all directions—each wave has a different polarization. By contrast, a *plane polarized beam* has the electric component of every light wave lined up in parallel order—each pointing in the same direction.

A polarizing filter is an optical component that preferentially transmits (lets pass) light of a specific polarization. If you view an ordinary light bulb through a polarizing filter you'll note some decrease in brightness, but not much more. Next time, however, view the light through a pair of filters: As you turn one filter in relation to the other you will see the bulb's brightness diminish to a minimum level at a specific orientation of the two filters.

The explanation is illustrated in Figure 5-9. The light bulb produces randomly polarized light; the first filter, though, blocks the transmission of light waves whose electric components are not pointing parallel to the filter's molecular structure. As you rotate the second filter you orient it so that it blocks all light waves that were passed by the first filter.

Arrange your diverging lens, viewing screen, and one of the polarizers as shown in Figure 5-10. Observe how the brightness of the projected spot changes slowly with time. This happens because the laser tube—by virtue of its design—produces plane polarized light, but the specific direction of polarization is random, and changes with time.

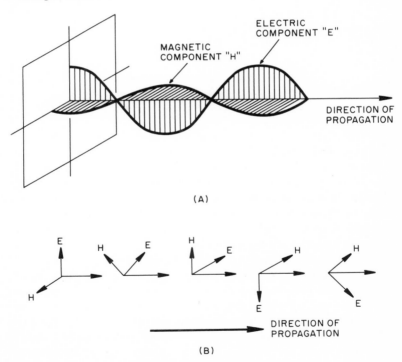

Fig. 5-8. Components of a light wave.

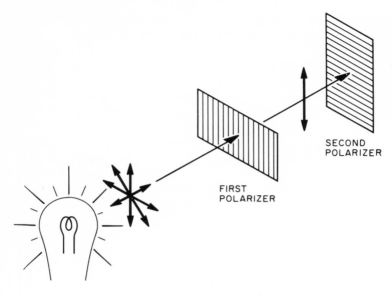

Fig. 5-9. How polarizing filters affect light.

Experiment #6: Single-Slit Diffraction

Diffraction is one of those phenomena that seems to go against common sense. After all, your common sense tells you that a sharp edge placed in a beam of light will cast a perfectly sharp shadow.

Sorry . . . but that just isn't true! The "sharpest" shadow is actually slightly fuzzy due to diffraction. The laser beam makes it easy to study diffraction phenomena by making them readily visible—the beam's monochromaticity and high intensity, coupled with its small diameter, are the reasons.

The reason that no shadow can be perfectly sharp is that light *bends* (or diffracts) around the edge, much like the ripples in a pond bend around plant stalks in the water. Moreover, when you examine the fuzzy part of the shadow (as you will, shortly), you can see a *diffraction pattern* of alternating light and dark lines. The famous seventeenth century mathematician, Christian Huygens, explained this phenomena by formulating his classic "principle:" When light emerges past a barrier edge, each point of the *wavefront* (the imaginary surface that is perpendicular to the direction of motion of the wave) acts like a new source of light. These "new" light waves— each "produced" by a "new" point source of light at the barrier edge—*interfere* with each other, and produce the classic diffraction patterns.

We must backtrack for a paragraph, and say a few words about *interference:* This is a phenomenon that can be observed in all kinds of wave motion, including electromagnetic wave motion. As we've said earlier, a light wave has both electric and magnetic components. These components are continuously varying in *amplitude* (or strength); the variation of either component can be plotted on a graph as a function of the distance the wave has traveled, and the result is a familiar sine wave (Fig. 5-11A). The plot shows a single light wave "frozen" in space: the X-axis represents the direction of *propagation* (or movement) of the wave; the Y-axis the amplitude of the electric component. Note that the amplitude smoothly rises to a

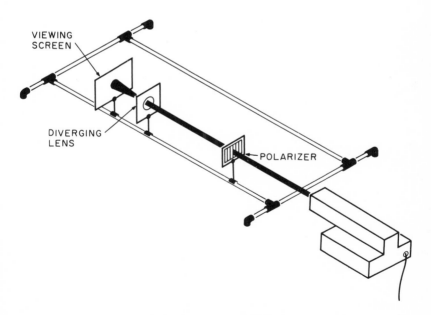

Fig. 5-10. Setup for Experiment 5.

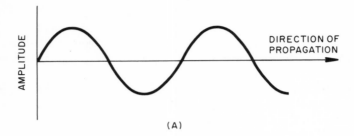

(A)

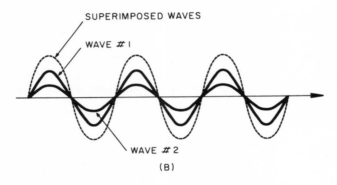

(B)

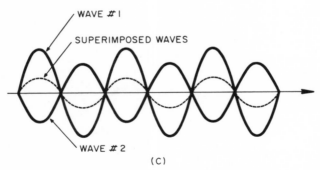

(C)

Fig. 5-11. Superimposed light waves: (A) a single light wave;
(B) constructive interference; (C) destructive interference.

maximum positive value, then sinks to a maximum negative value, as
it runs through the cycle again and again.

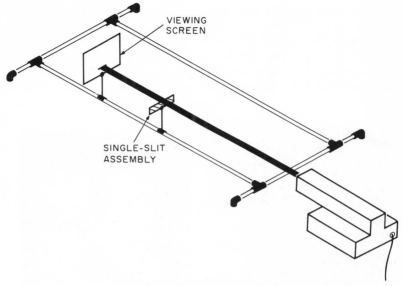

VIEWING
SCREEN

SINGLE-SLIT
ASSEMBLY

Fig. 5-12.   Setup for Experiment 6.

When two (or more) light waves come together, they literally superimpose: the various components add together to create a total component. Thus the electric components add together, and the magnetic components add together. If the light waves have the same wavelength (if they are the same color light), it is possible for the waves to actually cancel each other out. This is called *destructive interference:* it happens when the positive components of one wave overlap the negative components of the other (Fig. 5-11B). Similarly it is also possible for the two waves to merge together into a stronger wave; this is called *constructive interference* (Fig. 5-11C).

The various diffraction phenomena are caused by interference on a larger scale; holography is based on interference on a still larger scale, across the surface of a piece of film. The dark lines in a diffraction pattern are caused by the destructive interference of light waves; the bright lines by constructive interference. Similarly . . . the complex line and sworl pattern within the emulsion of a hologram is the captured *interference fringe* pattern caused by countless constructive and destructive interference "wave mergers" on the emulsion's surface.

Back to single-slit diffraction: The phenomenon gets its name from the fact that the light beam is diffracted by the two edges of a very narrow slit. For this experiment you can use either the soot-covered slide (with single slit) described in Chapter 3, or you can

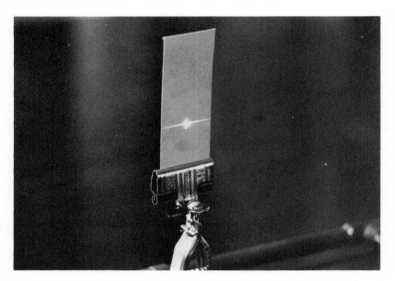

Fig. 5-13. Actual single-slit diffraction pattern that should be obtained.

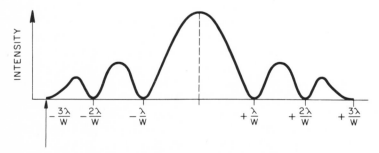

Fig. 5-14. Expected intensity pattern for single-slit diffraction.

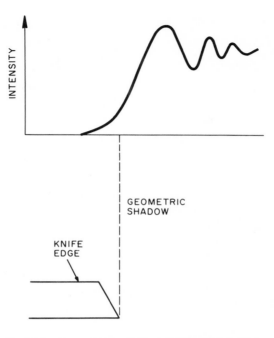

Fig. 5-15. Expected intensity pattern for diffraction by a straight edge.

make a slit by butting two single-edge razor blades together in a carrier-clamp, edge-to-edge, leaving a tiny, parallel-edged gap between them. Note that the narrower the slit, the wider the various visible fringes in the pattern will be.

Set up the components as shown in Figure 5-12. The beam must strike the slit squarely in the center. The visible pattern will increase in size—but decrease in brightness—as the viewing screen is moved further away from the slit: Select the distance for best viewing. The photograph in Figure 5-13 is of an actual single-slit diffraction pattern produced by the apparatus described in this book; the graph in Figure 5-14 is a plot of the expected intensity pattern on the screen. Note that the fringe spacing is determined by λ (the wavelength of the light) and W (the width of the slit). Adjust the slot to different widths to verify this concept.

Experiment #7: Diffraction by a Straight Edge

Replace the single-slit component with a single-edge razor blade; adjust the blade's position so that it cuts the beam halfway through. The resulting diffraction pattern is somewhat harder to see than the single-slit pattern because it does not have dark, zero-intensity bands. The graph in Figure 5-15 is a plot of the expected pattern; the spacing between maximum and minimum intensity points is determined by the wavelength of the light.

Experiment #8: Double-Slit Diffraction

Repeat the diffraction experiment using the double-slit component; adjust the slide's position so that the laser beam overlaps both slits. The pattern will look like the single-slit pattern until you examine the various bright regions. Each bright zone actually is made up of several fringes. *Note:* If you do not duplicate the type of pattern shown in Figure 5-16, refabricate the double-slit component (as described in Chapter 3). For best results, it is imperative that the slits be narrow *and* very close together.

Experiment #9: Diffraction by a Grating

Repeat the diffraction experiment using the diffraction grating slide; positioning is not critical, since the multiple slit pattern is uniform across the slide's surface.

The resulting diffraction pattern will look somewhat similar to the double-slit pattern, although much brighter. This is because diffraction by many adjacent slits produces a pattern that is com-

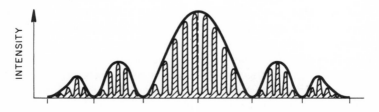

Fig. 5-16. Expected intensity pattern for double-slit diffraction.

Fig. 5-17.   General view of Interferometer setup.

posed of the superposition of many single-slit patterns. The result is a group of bright regions, each composed of several fringes. For detailed study, try projecting the pattern on a blank wall or a movie screen.

Experiment #10: Michelson Interferometer

With the exception of making and viewing holograms, this is perhaps the most exciting optical experiment you can perform. It is a classic demonstration that light waves will interfere with each other, and it dramatically illustrates the incredibly small dimensions with which we deal when we experiment with light. The interferometer, itself, has many practical scientific and industrial applications: It can be used to measure distances and dimensions to accuracies of a millionth of an inch, for example, or to provide unbelievably sensitive vibration measurements (as you will discover, shortly).

Figure 5-17 is a photograph of the interferometer setup; Figure 5-18 is a sketch of the component arrangement. Essentially

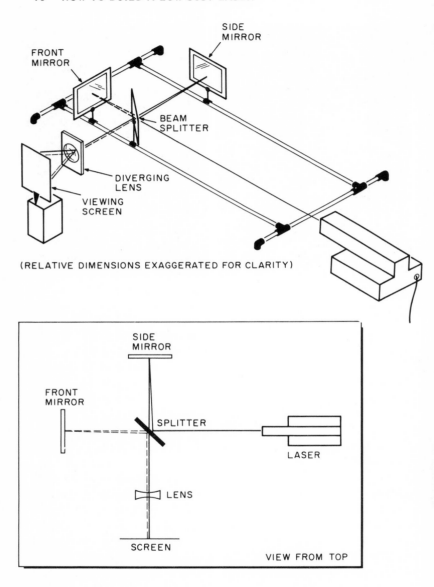

(RELATIVE DIMENSIONS EXAGGERATED FOR CLARITY)

Fig. 5-18. Setup for Experiment 10.

the laser beam is divided into two parts (by the beam splitter); each part travels along its own path for a short distance, then they are brought together and projected on a viewing screen by a single diverging lens. The beams interfere on the surface of the screen to create an alternating bright stripe/dark stripe pattern of interference fringes.

A solid base is absolutely critical, as is a closed room: The *slightest* motion of any component, or moving air currents in the beam paths, affect the relative *phases* of the beams, and set the pattern into motion, jumbling the fringes together.

Mount the beam splitter/lens assembly first. The carrier-clamp must be perpendicular to the rail; the splitter must make a 45-degree angle with the rail, when viewed from above. Do not tighten the lens thumbscrew at this time.

Next position the side mirror. Adjust the tripod-head clamp so that the mirror is parallel to the rail and perfectly vertical. Slide the carrier-clamp along the rail until the mirror reflects the side-ways laser beam back through the beam splitter; the paths to and from the splitter should be superimposed. Now adjust the position of the diverging lens so that the beam reflected back from the side mirror produces a round spot (about the size of a half-dollar) on the viewing screen.

Lastly, manipulate the front mirror (without disturbing any other adjustments) to reflect the forward laser beam back at the splitter *and through* the diverging lens. This may require that you angle the mirror *slightly*. (If you have difficulty with this step, recheck that both the beam splitter and side mirror are held perpendicular to their rails by their clamps; also be sure that the laser beam is parallel to the beam splitter rail.)

When the two beams are overlapping on the screen, stand back, and allow several seconds for vibration to die down. You should see the classic "venetian blind" interference fringe pattern. If you don't, the cause is probably excessive component vibration (caused by house movements much too small to feel); try the "floating bench" procedure outlined in the beginning of this chapter.

The pattern lines will probably slope to one side. Note how tiny changes in the position of the mirrors will change the angle.

And observe how sensitive the pattern is to air currents and/or movement . . . your interferometer will probably respond to a person walking in the next room!

The simplicity of the fringe pattern—as series of lines—is a result of the flatness of the two mirrors. By contrast, the fringe pattern within a hologram is far more complex: Here one of the beams bounces off the subject on its way to the film, and the subject has a complex surface. Note, however, the great similarity between the interferometer and the holography setup. The holography rig can be thought of as a special type of interferometer.

6

ALL ABOUT
HOLOGRAPHY

After you make your first hologram, and begin to show it off to your family and friends, you will discover a thorny problem: How to explain to them exactly what a hologram is.

Sure, it's convenient to say that a hologram is a "three-dimensional laser photograph," or that holography is "the science of lensless photography using laser beams," but both of these shortcut definitions are wrong. A hologram is not a photograph any more than a sculpture of a horse is a drawing of a horse; holography is not even a kissing cousin of photography—it's based on a totally different principle.

Even the most elaborate three-dimensional photograph—the kind you look at through a viewer—has an essential limitation: You view the subject from a single fixed point of view. It is as if you were wearing horse blinders that forced you to look at the subject from the exact spot that the camera was placed. This is the nature of photography because a piece of film placed in a camera is a *recording* device: It records the scene or subject from one specific point of view.

When you snap the camera's shutter, light waves bouncing off or produced by the subject *expose* the film. Specifically, these light waves have been gathered by the camera's lens and have converged on the surface of the film to create an image of the subject. The finished photograph you view is a record of this single image, locked

Fig. 6-1. A hologram looks like an underexposed photographic
negative.

forever in a pattern of metallic silver particles suspended in a gelatin
emulsion.

Remember the old saw that "a single photograph is worth a
thousand words?" Well, a single hologram is equivalent to an *infinity*
of ordinary photographs. That's because when a hologram is
"viewed," it reproduces the actual pattern of light waves that were
bouncing off the subject when the hologram was made. The precise
image you see depends on how you view the hologram: If you move
your head a tiny fraction of an inch, you will see a slightly different
view of the subject.

Suppose there is a hologram on the table in front of you. Pick
it up and examine it under ordinary room light. What you will see
(Fig. 6-1) is totally unimpressive: The hologram looks like an underexposed photographic negative—an almost transparent square of
plastic film covered with a few blotchy patterns of whorls and
sworls. These patterns have absolutely nothing to do with the
hologram; they are the accidental by-products of dust, dirt, and
optical imperfections in the hologram-making apparatus, so learn to
ignore them completely. In fact a very high quality hologram made
by professionals using rooms full of equipment is likely not to have

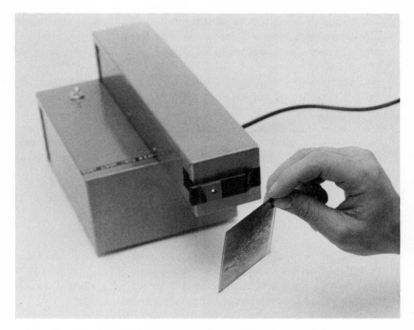

Fig. 6-2.   Locate the correct viewing angle to observe a hologram.

any such patterns; it will simply look like a square of transparent plastic.

To see what holography is all about, you'll need a source of coherent light—specifically, a laser equipped with a diverging lens to create a broad, low-intensity, coherent light beam. (The viewing set-up is described in Chapter 8, along with details on an incandescent light viewing technique that produces acceptable—if less than perfect—views).

Position the hologram so that the coherent light beam passes through the film at the same time that you look through the hologram back at the beam (it won't be very bright because it has been diverged), and juggle the hologram a bit to find the correct viewing angle (Fig. 6-2). You'll find yourself looking at an image of the subject that is as life-like as if you were peering through a window directly at the subject. The image you will see is fully three dimensional—but even more startling is the fact that as you move your head slightly you will see the subject from a slightly different point of view. You can actually see around, over, and under parts of the subject.

The two photographs in Figure 6-3 demonstrate this "parallax effect:" These are photographs of the holographic image made from two different viewpoints. Note that the perspective is different in each photo. Incidentally, the image quality you actually see when you view a hologram is far better than the quality depicted in the photo; it is difficult to photograph a holographic image created by a low-power laser.

History

As we've said earlier, the hologram/laser duo is actually re-creating the pattern of light waves that was reflected by the subject when the hologram was made. How this neat trick is done is one of the most exciting stories in contemporary physical science.

The story begins back in the late 1940's in Great Britian. Dr. Dennis Gabor, a Hungarian-born physicist working at the Imperial College of Science and Technology in London, was searching for a way to improve the performance of electron microscopes used to provide extreme magnification of submicroscopic subjects (such as viruses and the crystal structure in metals). Gabor deduced the basic hologram-making process (which we will discuss shortly) as a kind of adjunct to microscopy—as a possible way to greatly enhance the visible detail in micrographs—and in 1949 he published a now-classic paper on the subject. Dr. Gabor coined the term "holography" by combining the Greek words for "whole" and "image," but he was unable to produce true holograms because of the lack of a source of coherent light.

The invention of the laser filled this need, and in 1963 two researchers at the University of Michigan, Emmett Leith and Juris Upatnicks, made the first hologram. In 1971 Dr. Gabor was awarded the Nobel Prize in Physics for his pioneering work—recognition of the fantastic potential of holography in many diverse fields:

Fig. 6-3. Photographs of a single holographic image made from
two different angles.

Mechanical Design

Stress holograms, a kind of holographic double-exposure we will describe later, are used to help designers to locate points of maximum stress—and thus maximum wear and tear—in mechanical components for engines and large machinery.

Computer Technology

A single hologram actually "stores" an incredibly large amount of information (after all, one hologram is equal to countless separate photographic images), so computer designers are at work crafting holographic memory banks that will someday replace punched cards and magnetic tapes in many computer applications. Here, data will first be displayed in visual form—perhaps a pattern of white and black squares on a display board. Then a hologram will be made of the display board. For reasons we will cover shortly, a tiny-sized hologram is much better at doing this job than a conventional photograph. A special machine-controlled viewer would recall the stored data whenever needed.

High-Speed "Photography"

The hologram's ability to capture a true three-dimensional view of the subject with a single exposure can be utilized to "photograph" transient events such as small explosions, or the biological activities of short-lived specimens on a microscope slide, or the fracture of a stressed machine part. A high-powered ruby (or other crystal) laser that delivers a short, intense burst of coherent light is used as the light source. The investigator can study the event at his leisure; the hologram even makes it possible to measure parts of the subject in any dimension, and the image can be studied under a microscope. Because no lenses were used to make the hologram, all parts of the image are in perfect focus, and the detail captured is much finer than any photograph can display.

Display and Image Projection

The potential applications here are surprising. Designers are creating holographic traffic signs for roadside use that will display different messages when viewed from different angles; RCA scientists are working on a method of storing TV pictures on a movie-film-like strip of holograms. This gadget may provide an inexpensive way of recording TV shows for playback in your home, much as a phonograph record lets you play back music. Advertising men are contemplating the use of holograms in visual displays and outdoor advertising signs. And one company—McDonnell Douglas Electronics, in St. Charles, Missouri—currently offers miniature holograms that mount on business cards, and can be viewed with ordinary white light from a pen-cell flashlight. These are only some of the far-flung applications of holography.

When Light Wave Meets Light Wave

To understand how holography works, we've got to get down to basics. Let's start with a simple object—say a chess piece—sitting on a table. If we illuminate the chess piece with a beam of light, our eyes will see it on the table. What exactly does the phrase "Our eyes will see it" mean? Actually, it tells us that the chess piece reflects some of the light beam back at our eyes. Specifically, each tiny surface element of the chess piece reflects a bit of light back toward us. The light waves from each point join together to create an overall reflected light pattern that scientists call the *object wave*.

If you take a photograph of the chess piece, your camera's lens will transform the object wave into an image on the film's surface, and the film records this image. Holography, though, captures the object wave *itself* on a piece of film—the resulting hologram contains all the information necessary to recreate the complex jumble of light waves bouncing off the chess piece. When you view the hologram in laser light, the original object wave is duplicated, and you see exactly what you saw when you viewed the chess piece sitting on the table.

The apparatus used to accomplish this optical legerdemain is sketched in Figure 6-4. The laser is a source of coherent light; it will be clear in a few paragraphs why the illuminating source must produce coherent radiation. The laser beam travels to the *beam splitter* where it is divided into two components—each coherent.

The twin beams are then spread into cone-shaped beams by a pair of diverging lens. One beam—the *subject beam*—illuminated the subject, the second beam—called the *reference beam*—is directed at

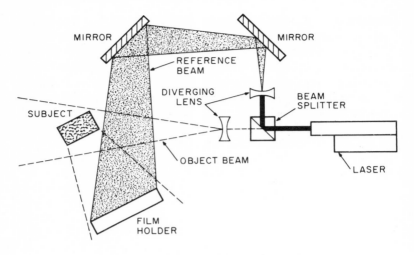

Fig. 6-4.   Setup geometry for making holograms.

the piece of photographic film (that will become the hologram) by a simple mirror system.

You'll note that light from two "sources" strikes the film's sensitive surface:

1. Light that reflects off the subject
2. The reference beam

If this were ordinary light, nothing much would happen (besides the fact that the film would slowly fog). *But*—and this is a *huge* but—the light from both "sources" is *coherent,* that is, *all* of the light waves striking the film are *monochromatic* (all have the same wavelength) and *in phase.*

As a result, the reflected light and the reference beam interfere with each other on the film's surface. A fantastically complex pattern of twisted, curving, and irregular interference fringes is formed on the surface. The spacing between the *maxima* and/or *minima* is much less than the neatly defined bright and dark line patterns with which we experimented in Chapter 5, so the pattern is not visible to the eye. However, the film's sensitive emulsion does respond to it to create, if you like, a "negative of the interference pattern." This, when it is developed, is the hologram.

It is important to note here that the reason the interference-fringe pattern is so complex is that the light waves reflected off the subject have become an extremely complicated—but still coherent—jumble. Clearly, each point on the subject, by virtue of its being a different distance from the light source, introduces a "phase shift;" this is a scientific way of saying that a given light wave will strike the front of the subject before it reaches the side. The resulting *object wave* is thus a composite of these slightly phase-shifted light waves; each tiny element of the subject provides its own contribution.

It takes a page of calculation to prove the next statement, so we will simply state it: *The "captured" interference fringe pattern on the developed film is actually a unique kind of diffraction grating.* You'll recall that the diffraction grating we used in Chapter 5 consisted of a pattern of thousands of vertical "bars" and "spaces." It produced the characteristic diffraction pattern we demonstrated in Chapter 5.

When we view a hologram, most of the viewing laser light passes through, but some is diffracted by the fringe pattern. This diffracted light is what we look at—it produces the image we see. In effect the fringe pattern transforms the simple viewing light beam into a replica of the original object wave. The hologram fringe pattern is, of course, much different than the ordered pattern of an experimental diffraction grating. The width of the "slits" varies from point to point on the hologram, as does the *contrast* of the non-transparent areas (their degree of non-transparency).

Holographic scientists explain the complete process by noting that the hologram making and viewing procedures are mathematical reverses of one another. The arithmetic that proves the point is tedious, but the heart of the matter is straightforward to explain: We construct a hologram by letting a complex light wave (the object wave) interfere with a reference beam on the surface of a piece of film. This creates a complex fringe pattern that is capable of acting like a unique diffraction grating—a kind of "optical template"—that will reconstruct the original object wave when illuminated by a viewing beam (actually this viewing beam is identical to the reference beam).

The fringe pattern on the hologram is the common denominator: It was created by the object beam *and* it has the ability to "distort" the viewing beam back into a replica of the object beam. As you would imagine, the fringe pattern created when a complex subject—like our chess piece—is hologramed is far too jumbled to analyze visually under a microscope.

Far-Out Phenomena

We've said that the image you see when you view a hologram is realistic. Actually, it is so realistic that it is impossible to come up with a visual test that will distinguish between the holographic image and the real subject (when viewed under identical lighting conditions).

Two reasons for this realism that we have already explained are the three-dimensional and the parallax effects; a third is the fact that your eye actually refocuses when it looks at near and far parts of the viewed image. All of these effects demonstrate that the original object wave is being reconstructed.

After you have made a few holograms you are ready to perform a really amazing trick: Take a hologram and cut it in half. Believe it or not, each half will show you a *complete* view of the holographed subject! Cut each half into quarters ... eighths ... sixteenths—each piece will still display the complete subject from a slightly different point of view.

This earthworm-like trick is the happy result of a fundamental principle of holography: *Every* tiny element of the film's surface receives light waves that are reflected from *every* point of the subject within its "field of view." Thus each tiny bit of film surface contains a complete hologram of the subject, as seen from its specific point of view.

It is this hologram characteristic that interests the designers of computer memories: A relatively small chunk of film can store a great deal of data. Holography is potentially superior to microfilm techniques in this application because it uses no lenses—and lenses, because of optical imperfections and *aberrations,* limit the ultimate performance of all photographic processes.

Another unusual talent of the holographic process is its ability to record more than one hologram on a single piece of film. This is not magic; it is a result of the fact that the angular orientation of the interference fringes helps to determine the apparent position of the image you see. Thus, after you have exposed a piece of film once, you can rotate it 180 degrees in the film holder and expose a second hologram. When you view the developed film you will see two different images, depending on how you orient the hologram.

Special Holographic Techniques

The basic techniques of holography have been expanded in recent years beyond the relatively simple procedures we will discuss in Chapter 8. All of the advanced methods require equipment and facilities found only in well-stocked and well-staffed optical laboratories. However, although we can't duplicate the results in our workshops, the following three holographic advances are important enough to discuss:

Stress Holography

The concept is simple, but the execution requires massive optical benches and the use of glass photographic "plates" instead of film. The technique involves making a double-exposure hologram of the same subject on a single piece of film (both subject and film must be maintained in precisely the same position for both exposures; hence, the requirement of heavy optical benches and glass film plates which won't buckle or flex with temperature change).

The subjects are typically structural elements (such as metal beams) or key components in machinery. The first exposure captures the subject in its "normal" state; the second is made with the subject stressed or heated. For example, a metal beam will be stressed to duplicate its usual carrying load, or a machine part will be heated (and possibly stressed, also) to duplicate its usual working conditions.

The resulting hologram shows the subject clearly, but, in addition, contains a superimposed pattern of stress lines that provides data on whether the component is properly designed to withstand the intended load. These stress lines (see Fig. 6-5) are present

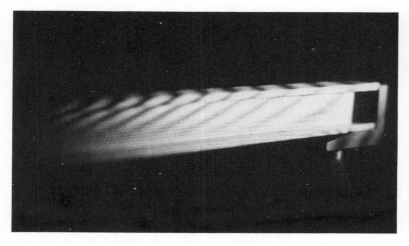

Fig. 6-5.   Hologram showing stress lines in a machine part.

because the dimensions of the subject change very slightly as it deforms under stress or heat. Excessive deformation, and hence poor component design, shows up as a tight, non-uniform stress line pattern.

Mechanical engineers call the technique *holographic interferometry,* and it probably will become a major design tool of the 1970's, especially when reliability of a design is a critical requirement.

Color Holography

This is a very involved process using three individual lasers— one with a red beam, one with a blue beam, and one with a green beam—that produces three holograms of the same subject on a single piece of film. When viewed (again, the viewing apparatus produces three coherent beams of different color) the viewer sees a "full-color" image. The principle of using three "primary" colors (red, blue, and green) to produce an image that *appears* to be full color is exactly the same technique used to create a color picture in a TV set.

Real-Image Holography

This technique will be broadly used in the future in a variety of consumer applications, such as advertising displays, special signs, and educational models. Essentially the technique produces a special hologram that has two unusual properties:

1. The reconstructed image appears *in front* of the hologram.
2. The hologram can be viewed in a beam of ordinary white light (from a simple penlight illuminator).

In optics terminology the image of a conventional hologram (the kind we will produce) is called a *virtual* image. The image seems to be behind the hologram even though the reconstructed object wave exists only in front of the hologram. By contrast, everything happens in front of a real-image hologram; light from the front white-light source causes the creation of a real image as it bounces off the hologram.

A real-image hologram is a sort of "hologram of a hologram." Both sides of the photographic emulsion are exposed simultaneously—one by the reference beam, the other by the image generated as a "readout beam" passes through a "master hologram" (a hologram of the subject).

The interaction of the dual exposing sources creates a very complicated fringe pattern that, besides recording the subject's image, acts like a color filter. This is why a white-light source can be used to view the hologram—all light wave "colors" other than laser red are automatically filtered out.

7

LASER PHOTOGRAPHY

Holography is radically different from conventional photography, yet it uses the basic materials and processes of photography, many of which are undoubtedly familiar to you.

A Few Words about Film

Holograms are recorded on *ultra-fine-grain* photographic film, film that is capable of capturing the microscopically fine detail of the pattern of interference fringes caused by the interaction of the two light beams (the subject beam, and the reference beam) on the film's light-sensitive surface.

The film you will use to make holograms was originally designed for specialized scientific applications. Although in some ways it behaves like ordinary snapshot film, it does have a few unusual idiosyncrasies. And since it is very expensive, it is important that you understand how to work with it.

Like most films, holographic film consists of a light-sensitive *emulsion* coated on a suitable *base* material—a thin transparent plastic. Professional holographers often use *photographic plates* made by coating the emulsion onto a thin glass sheet. Plates are more expensive than film, but offer the advantage of rigidity: Film flexing or waving can spoil a hologram.

The emulsion is a blend of special *gelatin* and tiny crystals of *silver bromide* and *silver iodide* (both these silver compounds are members of the family of chemicals known as *silver halides*). Although at first glance gelatin seems an unlikely material to be used for anything but cooking, it is actually a perfect choice for photographic emulsions. When heated it flows easily, so it can be readily applied to the base material. When cool it holds the silver halide crystals firmly in place. It is transparent, so light can strike the crystals. And it is relatively porous, so developing solutions can do their job after the film has been exposed.

When light strikes the silver halide crystals, they undergo minute internal changes that result in the formation of a *latent image*. Physicists do not yet understand the process completely, but it appears that photons (bundles of light energy) striking the crystals affect the attachment of silver atoms within the crystal structure. The latent image is invisible; "latent" means that the image is hidden, but nevertheless there. However, it can be made visible by placing the film in a suitable *developing solution*.

Developer is a blend of several chemicals that react *only* with those silver halide crystals that have been struck by photons. The developer changes these "exposed" crystals into particles of metallic silver. Because these particles are tiny, they look black. Unexposed crystals—those crystals that have not been light struck—are not affected by the developer.

The developed film is then placed in a *"stop bath"* solution which stops the development process after a specific period of time (which depends on the formulation of emulsion and developer, and on the developer temperature). Proper development time is a key element of all types of photography, and it is vital to creating perfect holograms.

Next the film is placed into a *fixer* solution, a chemical bath that dissolves away all those silver halide crystals which weren't exposed earlier. The fixer does not disturb the silver particles in the emulsion. At this point the film is no longer light sensitive.

Finally the film is washed in water to remove all traces of fixer solution (if not removed, these will stain the developed film) and then dried.

In conventional photography the film at this stage is called a *negative:* it has captured the image focused on its surface by the camera's lens. Of course the image is in reverse. Light areas in the subject have created dark regions within the negative, while dark parts of the subject have created clear, almost transparent, zones in the negative.

In holography the film at this stage is a completed hologram: Deep within the emulsion the complex interference fringe pattern has been "reproduced" as lines and bands of deposited metallic silver particles. Note that the actual hologram is invisible: A high-power microscope would be needed to see the silver patterns. However, a completed hologram often carries a pattern of visible lines and sworls. These are "recordings" of "optical noise" caused by dirt, dust, and imperfections in the optical components used to make the hologram. To view the hologram, you must use either the laser arrangement or the viewing device described in Chapter 8.

Obviously a piece of film can not differentiate between room light and the desired inteference pattern. Thus you must expose the hologram in total darkness. The only light striking the emulsion must come from the subject and reference beams; room light will *fog* the film (expose all the silver halide crystals). Processing also must be done in total darkness. Only after the fixing step—when all unexposed crystals have been washed out of the emulsion—can you turn on the lights.

Note that you can *partially* skirt the total darkness ban through the use of a *weak dark green* "safelight" (see the equipment list under *Darkroom Equipment*). *Do not use* a standard red or amber safelight: Holographic film is designed to be especially sensitive to red light, since a gas laser produces a red-colored beam. The glow from a red or amber safelight will fog the film instantly.

It's best, though, to learn to work in total darkness. The procedure—placing a sheet of film in place on the hologram apparatus, operating the "shutter" (simply a sheet of black paper), removing the film, and running it through the development process—is basically a series of simple, mechanical steps. Practice a few times while you are wearing a blindfold, and you will perfect it.

If you feel you must have a safelight, set it up so that it never comes closer than 4 feet to the film. Use it to light your path; not your hands.

The developer, stop bath, and fixer bath are all chemical solutions. And, although they are not particularly dangerous to work with *if handled properly*, they deserve your respect when you handle them. Precautions for working with the stop bath and fixer are clearly specified on the packages; the developer requires an extra word of caution because some people are sensitive to the solution, and will develop a skin irritation if they touch the solution. Although it is good procedure to handle the exposed film with tongs as you move it from solution to solution (this minimizes fingerprints on the soft emulsion surface), it's almost a law of nature that some part of your hand will come into contact with the solution sooner or later. Wash your hands off if this happens—and if you know you have sensitive skin, use rubber gloves when you work.

Similarly, plan to set up the film processing trays on a formica or other chemical-resistant surface. The solutions can stain wood finishes, and accidental spills are frequent (another law of nature!).

Materials and Equipment

Film

Film for holography must have exceptional resolution capabilities, since it will be called upon to record microscopic patterns. Ordinary fine-grain films (such as Kodak Panatomic-X and others) will *not* work.

The only readily available suitable film is Agfa 10E75, manufactured by Agfa-Gevaert Inc., Scientific Products Department, 275 North Street, Teterboro, N. J. 07608. The film is supplied in large—and expensive—rolls by the manufacturer. You can purchase the same film in small quantities, precut into 75-mm squares from Edmund Scientific Co., 300 Edscorp Building, Barrington, N. J. 08007; or from Metrologic Instruments, Inc., 143 Harding Avenue, Bellmawr, N. J. 08030.

Chemicals

Developer solution is less difficult to obtain. The recommended developers are Agfa Metinol U (a solution normally used for developing photographic paper), and Kodak D-19 (normally intended to be used with special scientific film). A well-stocked photoequipment store is likely to have Metinol U on hand, and can usually order D-19. In a pinch you can use almost any developer formulated for fine-grain, high-contrast photography.

The two firms listed above supply a packet of Metinol U with each order of holographic film, so we will describe the hologram-making technique in Chapter 8 in terms of the use of Metinol U. Keep in mind that other developers may require a different development time to work properly. If you select a different solution, consult the manufacturer's instruction sheet to determine ideal development time.

Stop bath is totally non-critical. Any conventional acid stop bath can be used.

Fixer should be a *hardening* type solution that will harden the soft gelatin emulsion as it fixes the hologram. Again, any fixer designed for *film* can be used.

Note that *all* photographic chemicals are sold in either powdered or concentrated liquid form. Metinol U comes as two envelopes of powder that you mix together with water. Stop baths and fixers are available both in powder and concentrated liquid form (you dilute a concentrated liquid product to produce a "working solution"). It is crucial that you follow manufacturer's instruction *to the letter* when you mix or dilute the products, especially in regard to the temperature of the water used.

Darkroom Equipment

Darkroom equipment requirements are fairly modest because you will be using small squares of film, and because you will only process one hologram at a time. The list below spells out the gear you must have, as well as some optional equipment you might want to have:

1. Four small developing trays—inexpensive plastic or hard rubber trays that are available at any photo equipment shop. The small size (4 X 5 in.) is ideal. Three of the trays will be used to hold the three processing solutions; the fourth will serve as a "washing tank" to soak the completed holograms in running water.

2. Darkroom solution thermometer—this is a must! An inexpensive glass thermometer is acceptable, although many people find the red liquid column hard to read. If you can afford it, buy a dial-reading, metal-case thermometer.

3. A pair of film tongs—this is a minimum requirement; for better darkroom manipulation, buy two pairs of tongs. You should always use tongs to transfer the film from one solution tray to the next. The tongs keep your hands out of the solution, and minimize fingerprint marks on the soft emulsion. But, if you have only one pair, you must remember to wash them off in running water after *each* transfer. This step is necessary to avoid contaminating the developer solution with either fixer or stop bath. If you buy two pairs of tongs, use one to lift film out of the developer tray and drop into the stop bath (don't let the tongs touch the stop bath), and use the other to move film between the stop bath, fixer, and washing trays. Contamination is not a problem here.

4. Measuring cup—an ordinary 1-quart glass or plastic measuring cup designed for cooking is fine (but be sure to scrub away all traces of chemicals before you return it to the kitchen).

5. Storage jugs—you will need two or three plastic or glass jugs to hold mixed solutions. They should be dark-colored to minimize exposure of the solutions to light, which can cause deterioration.

6. Stirring rod—a commercial paddle-type stirrer works best, although almost any glass or plastic rod or implement will do the job. Do *not* use metal gadgets of any kind; some metals will react with the solutions and cause contamination.

7. Development timer (optional)—this is a useful device for timing the film's stay in the developing tray. Although less convenient, you can use a watch or clock equipped with a luminous dial and a sweep-second hand, or you can have an assistant outside the darkroom keep track of development time.

8. Safelight (optional)—an inexpensive model using a 7½-watt bulb (such as the Kodak "Brownie" Safelight) is perfectly acceptable. Remember, the safelight must have a *green* filter.

9. Rubber gloves (optional).

Darkroom Technique

The development process consists of a series of chemical reactions. Actually, the chemicals in the three processing solutions do all the work automatically. Your job is simply to create the proper conditions for the reactions to take place.

Temperature

Probably the single most important condition is temperature. Virtually all black-and-white developing solutions, including Metinol U, stop bath, and fixer, are formulated to work best at 68°F. Reactions occur more slowly at lower temperatures, and more quickly at higher temperatures, but quality may suffer. Therefore, do everything you can to maintain processing solution temperature at (or very close to) 68°F.

If your darkroom temperature is higher or lower than 68°F, fill a large shallow tray with 68°F tap water (use your thermometer to adjust the hot and cold faucets to create a 68°F flow). Then set the three solution trays into this "fixed temperature bath." The large mass of water in the shallow tray will maintain solution temperature at the correct point for 30 minutes or longer, enough time to process two or three holograms.

Washing can be done in *cool* water; temperature is not critical for this step.

Agitation

Another requirement is *moderate agitation.* Some film movement within the processing trays is desirable to insure that fresh solution continuously washes across the emulsion surface. You can accomplish this by *gently* rocking the tray every 20 or 30 seconds, and setting up a *slight* sloshing motion.

Solution Efficacy

Finally there's the need to maintain solution efficacy—the ability of each solution to do its job. The solution chemicals deteriorate with use, as they react with the emulsion silver halide crystals. If you use small trays (4 × 5 in.), refill the trays with fresh solution each time you process five holograms, or at the start of a new day's holography (never save a tray full of solution for more than a few hours—exposure to air can deteriorate the solution). Larger trays, holding a larger volume of solution, will process proportionately more holograms.

Keep the solution storage jugs tightly stoppered to protect the solutions from the air. In any case, discard any unused developer solution that is more than two months old (unmixed packaged developer keeps indefinitely.)

And be sure that you don't accidentally contaminate developer solution by placing in it a stirrer or thermometer that has just been in stop bath or fixer solution. Rinse off *all* utensils after *each* use.

Physical Layout

Most people find that it is best to position the three solution trays in a horizontal line: developer on the left, stop bath in the center, and fixer on the right. The actual development procedure is simple. Place the film first into developer solution, then into stop bath, next into fixer, and finally under running water. We'll cover the correct times for each stage in Chapter 8.

The washing tray simply sits in a convenient sink, directly under the spout. Adjust the faucets to produce a steady trickle of water into the tray that continuously pours fresh water into the overflowing tray. The procedure for drying the washed film will be covered in Chapter 8.

Handling the Film

Fine-grain emulsions are very thin, and at first you probably will find it difficult to determine in the dark which side of the film must face the front of the holder (the emulsion), and which side faces rear (the base). The trick is to educate your index finger: The emulsion has a creamy-smooth feeling, while the base has the slick-smooth feeling of a polished plastic surface.

The best idea is to sacrifice a single sheet of film, and use it as a "training device." Take the package of film into a totally dark room, remove the inner protective package and unfold the seal, and take out one square of film. When you have reclosed the film pack, turn on the room lights. Almost instantly the piece of unprotected film will turn a bluish-gray. Actually only the emulsion is changing colors, due to the effect of light on the silver halide crystals.

Examine the film carefully under a bright light; you'll be able to see which is the emulsion side and which is the base side. Now gently run your index finger over both sides until you are sure that you can identify the emulsion side by touch.

8

MAKING AND VIEWING HOLOGRAMS

Making a viewable hologram is no more difficult, technically, than making a conventional photograph . . . with one important difference: Holography doesn't tolerate "minor" mistakes.

If you jiggle a conventional camera, you'll still get a picture—it will be blurred, but it will still be there. Jiggle a holography apparatus, though, and you will end up with a "blank" hologram. Similarly, exposure times and development times are slightly more critical, and have an important effect on the quality of the image you will see.

There is an excellent chance that the very first hologram you make will be a good one, especially if you take the time to verify that your work surface is rock-stable, and you work a bit at mastering ideal "shutter technique." And so, expect success, but don't despair if your first attempts don't succeed. Remember that holography is a fresh-from-the-laboratory technology. The techniques and equipment are new to you, and it may take some practice to master them. Above all, think positive!

Begin by selecting a stable work surface: Relative motions among the various optical components and the film emulsion surface of even a few millionths of an inch will ruin the hologram. I cannot stress the importance of this factor enough . . . it will probably be the cause of most (if any) of your ruined films.

Follow the suggestions given in the beginning of Chapter 5 for locating a vibration-free work surface; verify the degree of stability by setting up the interferometer (Experiment #10) and observing the fringe pattern. *The pattern must appear stable—with no visible bar movement—for at least 20 seconds.* Fans should be turned off, air ducts closed, and windows shut, in order to minimize room air currents. And before you examine the pattern, wait several seconds for the air currents your body movement generated to die down.

Equally important is the requirement of total darkness (except for the presence of a *green* safelight, if you wish). Total darkness means just that: Typically, the piece of sensitive film will be out in the open for at least 1 minute—maybe even longer if you lack nimble fingers. Thus it is vital that your work area and developing dark room be totally free of *any* red or blue light (or white light, since it contains both red and blue components) other than the laser beam.

Use your eyes as a "test instrument:" Allow 2 full minutes for your eyes to get accustomed to the darkness, and then scan the room carefully. *Any* light that you can see (through cracks and crevices, under doors, past window shades) will probably fog the film.

Clearly, if room darkening is a problem, plan to perform all holography experiments at night. And consider purchasing inexpensive black photographic "drop cloths" to use as room darkeners on windows and doorways.

Setting Up for Holography

The most familiar holography setup is sketched in Figure 8-1. (Also see Fig. 6-4.) There are other possible geometries, including configurations that use only one mirror. However, the geometry shown is simple, and proven, and it is the one I recommend you use. Setting up is an eight-step procedure:

1. Note that the optical bench is turned sideways—with its long dimension perpendicular to the laser beam. Begin by mounting the beam splitter/diverging lens assembly on the rail nearest the laser. Be sure that the splitter is held vertically, and that it makes an angle of 45 degrees with the rail when viewed from above. Do not adjust the two diverging lenses at this time.

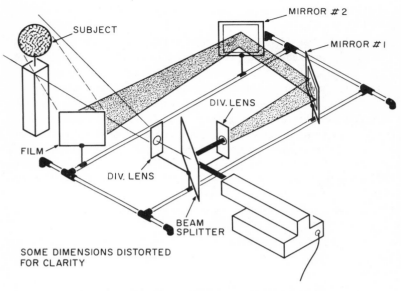

Fig. 8-1. Equipment setup for making holograms.

2. Mount the first mirror (mirror #1) in the location shown; the mirror must be positioned vertically, and it should make a 45-degree angle with the rail when viewed from above. Make sure the tripod-head adjuster is securely tightened.

3. Similarly, mount the second mirror (mirror #2), but don't tighten the adjuster more than finger tight; you will probably have to readjust this mirror later.

4. Put a piece of white cardboard in the film slot on the film holder (to act as a temporary viewing screen), and mount the holder on the bench, in the location shown. The holder must be held vertical; "point it" toward the final location of the subject.

5. Turn the laser on, and simultaneously adjust the side diverging lens, and the position of the second mirror (if necessary), to produce a large red area on the screen. Because the screen is angled with respect to the beam, the area will be shaped like a

football. Tighten the lens's thumbscrew and the mirror mount adjuster.

6. Place the subject stand in the location shown. Carefully adjust the front diverging lens to project an illuminating beam on the subject (Be sure that you don't upset other adjustments).

7. Examine the beam splitter, and you will see that there are actually two sideways beams (one from each surface), although only one is passing through the small diverging lens aperture. To prevent any detrimental effects from reflections of the second—unwanted—beam, carefully stick a tiny piece of black tape on the surface of the beam splitter to block the unused beam. *Note:* It makes no difference whether you use the front or rear surface beam.

8. Use a ruler to measure the two beam paths:

(i)   Splitter—mirror #1—mirror #2—film holder.

(ii)   Splitter—subject—film holder.

Ideally the paths should be equal (or close to equal). If necessary, adjust path length by sliding both mirror holder carrier-clamps toward or away from the center of the bench.

Another rule of thumb: The intensity of light reflected off the subject (the object beam) should be approximately one-third the intensity of the direct beam (the reference beam). Dim the room lights, and compare the two by alternately blocking each with your hand.

Complete the set-up by verifying that all thumbscrews, clamps, and adjusters are tight, and that all components are held in place securely.

Selecting a Subject

The combination of relatively low laser power and small film sheet size places a few obvious restraints on the variety of subjects you can choose:

• The subject must be physically small (no larger than about 2 in. long by 1 in. high by 1 in. deep).

- The subject must have a shiny, reflective surface. Avoid matte-finished or granular-finished subjects.
- The subject must be a good reflector of red light—which means that it should be colored white or red.

White chess pieces, small model cars and trains, small china figurines, small stones, and similar bric-a-brac are good candidates.

Mastering Shutter Technique

The laser beam must be allowed to stabilize for a full 30 minutes before it is used to make a hologram; thus the on-off switch can't be used to control the beam during the actual holographing operation—you must use a makeshift "shutter." The simplest and best is a sheet of black paper placed between the laser and the beam splitter; you lift the paper up to "expose" the hologram.

But there's a rub: You must develop a shutter technique that doesn't cause any of the components to vibrate, *and* that minimizes local air disturbances. The easiest, and most effective, way to do this is to practice shutter technique using the interferometer setup. You know you are ready to make holograms when you can "open" the shutter and produce a stable interference fringe pattern, with absolutely no fringe motion. Here is the technique I recommend:

The shutter is an ordinary piece of black "construction paper" (8½ X 11 in.). Block the beam by leaning the paper against the laser (with its bottom edge resting against the table).

Make an "exposure" (after the film has been placed in the holder) by performing the following steps in sequence:

1. Lift the paper off the table gently, but do not unblock the beam.
2. Hold the paper perfectly steady for at least 10 seconds, to give the bench time to quiet down from the "shock."
3. Take a medium-deep breath and hold it.
4. Carefully—and steadily—*slide* the paper out of the beam's path; it is vital that you don't stir up air currents by "fanning" the paper as you lift it. Pretend that the paper is a heavy door moving in a grooved slot.

5. Count out the appropriate exposure time to yourself.
6. *Slide* the paper back into place.

Exposing the Hologram

Allowing the laser to warm up for at least 30 minutes gives the tube a chance to stabilize. This helps insure that the output beam will be coherent for the total exposure duration. Use the warm-up time to examine the optical arrangement; look for stray reflections that may fall on the film holder (this can happen if the object beam grazes a shiny surface), and double-check that no part of the object beam strikes the second mirror (and is reflected toward the film holder).

When all is ready, darken the room completely and take one sheet of film from the double-wrapped light-tight package. Handle the film by the edges only: Do *not* touch the emulsion with your fingers (except at the edges when you feel for the emulsion side). Carefully place the film in the holder, *emulsion side facing the subject*. Make certain that you don't bend or buckle the film as you handle it, and try not to jar the optical bench, or disturb any adjustments, as you drop the film in place. Before you do anything else, double-check that the film package is tightly closed.

Allow about 30 seconds for the air currents you have generated to die down, then expose the film. I recommend a 2-second exposure for your first attempts; later you will want to experiment with shorter and longer times, to find optimum exposures for different subjects.

Developing the Hologram

Remove the *hologram* carefully (it is no longer simply a piece of film), and take it to the developing room. *Warning:* The total darkness requirement still is in force. A nearby bathroom is an obvious choice for a "darkroom." The three processing solutions described in Chapter 7 should be ready—and at proper temperature—in three processing trays.

Slide the piece of film into the *developer* solution, *emulsion side up*. Rock the tray gently to dislodge any air bubbles. Total development time is about 8 minutes; gently agitate the tray every 30 seconds (the purpose is to freshen the solution in contact with the emulsion).

Transfer the film to the *stop bath* tray; rock the tray gently for about 30 seconds.

Transfer the film to the *fixer* solution—once again, *emulsion side up;* rock the tray gently to dislodge air bubbles. Total fixing time is about 20 minutes, but you can turn *low-level* room lights on after the film has been in the fixer for a *full* minute. Do not short-change the fixing step: it is vital for the long-term life of your hologram.

Transfer the film to the *washing tray*. Wash the hologram for at least 20 minutes. This is necessary to remove all traces of fixer, and prevent the formation of stains. If you wish, you can shorten the washing process by dipping the piece of film into a commercial "fixer remover" solution (your local photo supply shop will stock it; follow the dilution instructions carefully).

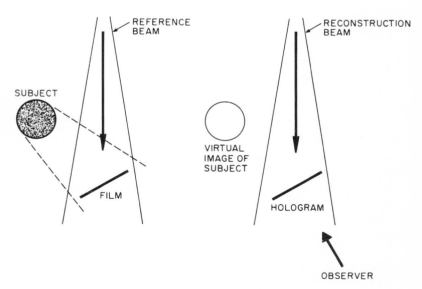

Fig. 8-2. Hologram image reconstruction.

Fig. 8-3. Lens assembly taped over laser beam hole.

Air-dry the film by suspending it in a *dust-free* room; use an inexpensive "film clip" to hold the hologram by its edge. The hologram will probably curl slightly as it dries.

Viewing the Hologram

It can be a bit tricky to learn how. The image reconstruction process requires that the reconstruction beam strikes the hologram from the same angle as the original reference beam (Fig. 8-2). The easiest way to create a reconstruction beam is to tape one of the diverging lens assemblies over the laser's beam hole (Fig. 8-3). This produces a perfectly-safe-to-view diverged beam of coherent, monochromatic light.

The hologram should be flat for optimum viewing; if excessive curl is a problem, mount the hologram in a metal or cardboard frame (mounts for 70-mm film can be ordered through many full-line photo supply shops).

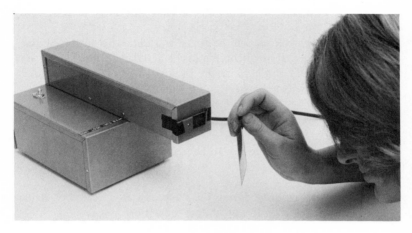

Fig. 8-4. Viewing a hologram.

Set the laser up on a table, darken the room lights, and position the hologram about 18 inches away from the diverging lens, *emulsion side facing forward.* Swing the hologram around until it intersects the beam at the same angle as the reference beam struck the film holder. Look through the back of the hologram; you will see the image of the subject as if you were looking through a window (Fig. 8-4).

Note: You may have to jiggle the position to find the best viewing angle. If you do not see any image, rotate the hologram 90 degrees (do not point the emulsion *away* from the beam; turn it as you would a wheel) and try again. Remember, the hologram must be oriented exactly as it was in the holder. Reversing the hologram (viewing it emulsion side rear) creates a "pseudoscopic" inside-out image that is impossible to describe.

There is a down-and-dirty way to view a hologram *without* a laser. You will need a red filter (an inexpensive gelatin filter available at most photo shops works fine) and a penlight-type flashlight; the penlight must have a point-source-type bulb (the kind with a tiny lens built into the tip). Place the filter directly behind the hologram, and hold the penlight about 12 inches behind the pair. The flashlight beam must strike the back of the filter at the same angle as the reference beam hit the film holder. View the image as described above.

The image you see will be fuzzy and somewhat distorted; this results from the non-monochromatic characteristic of the light (even after it is filtered) and because the filter's red color doesn't exactly duplicate the laser's red light.

If You Don't See Anything . . .

Don't panic! There has to be a simple explanation—probably one of the following:

- Something moved during the exposure. Recheck your stable base; practice shutter technique some more.
- The film was placed in the holder with the emulsion facing the rear. Work to develop an educated index finger.
- You made a developing error. Check that all solutions are mixed properly, and that you have processed the piece of film in the correct sequence.
- Exposure time was too short for your subject. Try a longer exposure.

INDEX